T0269309

Compact Textbooks in Mathematics

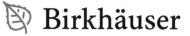

Compact Textbooks in Mathematics

This textbook series presents concise introductions to current topics in mathematics and mainly addresses advanced undergraduates and master students. The concept is to offer small books covering subject matter equivalent to 2- or 3-hour lectures or seminars which are also suitable for self-study. The books provide students and teachers with new perspectives and novel approaches. They feature examples and exercises to illustrate key concepts and applications of the theoretical contents. The series also includes textbooks specifically speaking to the needs of students from other disciplines such as physics, computer science, engineering, life sciences, finance.

More information about this series at http://www.springer.com/series/11225

Victor M. Panaretos

Statistics for Mathematicians

A Rigorous First Course

 Birkhäuser

Victor M. Panaretos
Institute of Mathematics
EPFL
Lausanne, Switzerland

ISSN 2296-4568   ISSN 2296-455X   (electronic)
Compact Textbooks in Mathematics
ISBN 978-3-319-28339-5   ISBN 978-3-319-28341-8   (eBook)
DOI 10.1007/978-3-319-28341-8

Library of Congress Control Number: 2016940379

Mathematics Subject Classification (2010): 62-XX

Printed on acid-free paper

This book is published under the trade name Birkhäuser
The registered company is Springer International Publishing AG Switzerland
(www.birkhauser-science.com)

In memory of David A. Freedman,
master of clarity

Preface

This book is intended as a text for Mathematics students taking their first course in Statistics, and grew out of my second-year course for mathematics undergraduates at EPFL. It is a book on "Statistics for Mathematicians" rather than on "Mathematical Statistics": the intent is not to focus on the deeper mathematical/theoretical aspects of the subject but rather to provide an introduction to the basic notions tailored to the mindset and tastes of the Mathematics student. Mathematics students are sometimes put off by the informal nature of first courses in Statistics, since many results are usually stated without proof or are accompanied by heuristic sketches of proofs. Another risk may be that of "intellectual entropy", when too many (and diverse) topics are covered in a single course, risking the impression of Statistics as a collection of recipes lacking natural connection. This book can be used as a basis for an elementary semester-long first course in Statistics that presents the basic ideas of one-parameter inference in a coherent manner, while making essentially no sacrifices on matters of rigour. It is meant to be compact, so as to be realistic to be covered in full during a single semester, and yet hopefully attract mathematics students to pursuing further elective courses in Statistics. In more detail, the three main tasks this text sets out to address are as follows.

(1) To provide a rigorous yet elementary course The effort is to prove essentially all the results rigorously. These results include some of the most central results such as the asymptotics of maximum likelihood, optimality in testing, asymptotics of likelihood ratio tests, and optimality results regarding confidence intervals. It also contains detailed proofs of some elementary results that are rarely worked out in detail in elementary texts (for instance, the derivation of the distribution of the t statistic). The only results not proven in the main text are some background results in probability and analysis. In the case of the probabilistic results, detailed proofs *are* in fact given in the appendix, and the proofs are still at an elementary level. These include results such as the continuous mapping theorem, Slutsky's theorem, the (third moment) central limit theorem, and results pertaining to moment generating functions. The analytic results not proven are Taylor's formula and the univariate inverse function theorem. These are stated in the appendix, where precise references are also provided for their proofs. In principle, thus, the course only requires students to have taken a first course in ϵ/δ-level analysis (including sequences, convergence, series, multivariable differentiation and Riemann integration, and Taylor's formula) and a first course in probability (including basic operations on events and the corresponding probability calculus, discrete and continuous random variables, joint/conditional/marginal distributions, and expectation/variance/covariance). A succinct fact sheet on all the probabilistic prerequisites is provided in the appendix, for easy reference.

(2) To provide a conceptually compact course, with a firm sense of direction The entire book can realistically be covered in full during the course of a semester, and it is also realistic for the students to solve all the exercises during the same period of study (a solution manual is available upon request for instructors). I have reduced the number of topics covered in order to be able to have the minimal number of topics that can be covered during a semester course without compromising on the mathematics, while still providing an overview of the main ideas of statistical inference. The course covers the basics of exponential families, exploratory data analysis, sampling, estimation, testing, and confidence intervals. It's true that the book does not tell the whole story and avoids detailed discussions of all the possible complications and variants in each section. However, I believe that the topics covered give a firm basis for the students to build on, and every attempt has been made for the story it tells to flow naturally, without giving the appearance of a collection of techniques. There is extensive cross-referencing of the material, illustrating how the different results are tied together, and an effort to develop the material in a "linear fashion", explaining why one is doing whatever they are doing at every point, and what the ultimate purpose is. No result is mentioned in vain (any results presented are subsequently used), and results are accompanied by substantial motivation and discussion. References made to results are always accompanied by the number of the said result, along with the page number in the book which allows for easy reference and self-study.

(3) To provide a course that is not on "Mathematical Statistics" but rather on "Statistics for Mathematicians" The audience is primarily intended to be undergraduate mathematicians, whom I hope to attract into Statistics rather than statisticians to whom I might want to introduce the more mathematical aspects of Statistics. Therefore, the course is not primarily intended to be a course in statistical theory. Rather, it is intended to be an entry-level course in statistical inference, presented in a way that would be more receptive by an audience comprised of mathematicians. Therefore, the discussion of different topics and the style and considerations are adapted to such an audience. For example, optimality, whenever discussed, is not presented as an end in itself but rather as a means of motivating methodology (the idea being that mathematicians would be motivated by "best" results more than by heuristics).

The means to balance the requirement of an elementary yet rigorous text was to adopt the use of the exponential family of distributions throughout (rather than aiming for full generality). This is of course a restriction, but in some ways not a major one, since most of the examples treated in elementary textbooks *are*, in fact, exponential families. Focusing on exponential families not only allows for elementary proofs using basic analysis and probability but also allows for the statements of the theorems and the required conditions to be simple and intuitive. Whenever results do hold more generally, this is remarked as a side note. A more detailed description of the structure of the text, and the progression of topics, can be found in the "Brief Overview" Section (p. 1).

The main concessions that regrettably had to be made in terms of coverage pertain to regression and the Bayesian paradigm, and this deserves an apology. The textbook is based on the first Statistics course that mathematics students take at EPFL, but this course is also the only *compulsory* course in Statistics. It may thus well be their last (though hopefully the book will convince them otherwise). In this case there is a dilemma. Does one strive to include as many topics as possible, so that the student be well equipped in the future in case this is all the Statistics they will ever see? Or does one try to cover a minimally sufficient number of topics as clearly and completely as possible, hoping that at least these topics will stick to mind? I opted for the second approach, as my impression is that adding more topics does not guarantee that these topics will in fact be remembered (in fact, a student having only taken a single Statistics course and finding themselves needing Statistics later will almost certainly have to do further reading anyway) and because this approach is more in line with the effort to produce a course with low conceptual entropy. For instance, notions such as p-values and confidence intervals are quite subtle to understand upon a first encounter (avoiding flawed interpretations such as "the probability that H_0 is valid" or "the probability that the parameter falls in the interval is 95%"). When the student does not already have a solid grasp, it may be unsettling—or worse still confusing—to suddenly switch things around.

In writing the book, and preparing examples and exercises, I have drawn inspiration from many excellent textbooks that have stood the test of time (but also more recent online resources, including Wikipedia and mathstackexchange). In doing this, I tried to balance the rigour found in advanced textbooks focusing on Mathematical Statistics, with the more accessible style of entry-level textbooks focusing on the basics of statistical inference. The former category includes Lehmann and Casella [15], Lehmann and Romano [16], Cox and Hinkley [6], Bickel and Doksum [1], Schervish [22], Shao [23], and Young and Smith [26], and the latter category includes Rice [19], Hogg and Tanis [13], Hogg and Craig [12], and Silvey [24] (the last one perhaps bordering with the first category). The book by Knight [14] strikes a very nice balance between the two objectives, though still at a level higher than the present text aims, and has also been an important source of inspiration and exercises/examples. More texts striking a good balance and including a more comprehensive list of topics than the present one (but still not including several proofs) include Casella and Berger [4], Davison [9], and Wasserman [25]. The necessary probability background for the present text is covered quite nicely in the first three chapters of Knight [14], but of course there are several texts devoted specifically to elementary probability (i.e. non-measure theoretic probability) that would suffice (e.g. Blitzstein and Hwang [3], Dalang and Conus [8] (in French), Grimmett and Welsh [11], Pitman [18], and Ross [20]). As mentioned earlier, Sect. A.1 contains a quick overview of the main prerequisites, for ease of reference.

While the main audience for the book will be instructors and students in mathematics undergraduate programmes, the textbook could still be used for programmes of study with substantial mathematical content, for instance, students of physics, economics, computer science, and engineering programmes looking for a more formal coverage of one-parameter inference. After all, to think like a mathematician is to think rigorously, regardless of the subject matter at hand.

In closing, I would like to express my gratitude to my PhD students and my undergraduate students whose meticulous comments and suggestions helped improve earlier drafts. Marie-Hélène Descary, Mikael Kuusela, Valentina Masarotto, Matthieu Simeoni, and Yoav Zemel provided extensive feedback, suggestions on exercises, and help with proofreading and layout. I especially enjoyed chatting with Yoav Zemel about how to best tiptoe around measure theory in the proofs of some more delicate results in the appendix (while remaining fully rigorous). I am also very thankful to two anonymous reviewers, who read a first version of the book and gave constructive and encouraging feedback. Any remaining glitches are, of course, my own. Finally, I would like to thank Veronika Rosteck and Springer/Birkhäuser for our pleasant collaboration.

Lausanne, Switzerland   Victor M. Panaretos
October 2015

Contents

Brief Overview

In a general sense, one can describe Statistics as the mathematical discipline whose purpose is to

use empirical data generated by a random phenomenon, in order to

make inferences about some deterministic characteristics of the phenomenon

while simultaneously quantifying the uncertainty inherent in these inferences.

Let's take a step back and consider the different elements of this description. What is a random phenomenon? We can think of a random phenomenon as a system or process whose outcome X is uncertain. This means that, even if we know every aspect of this system or process, we cannot perfectly predict its outcome X. Mathematically, such phenomena are formalised via the theory of probability: the outcome X is a random variable, and the model that describes the phenomenon is the probability distribution function $F(x) = \mathbb{P}[X \leq x]$ of this random variable. Now there may be a characteristic θ of this phenomenon that influences the probabilities associated with the outcome of X. Such a characteristic is called a parameter. Since the probability of $\{X \leq x\}$ is influenced by θ, the function $F(x)$ must be a function of θ, so we write it as $F(x; \theta) = \mathbb{P}_\theta[X \leq x]$.

If we know the functional form of $F(x; \theta)$, and the true value of θ, we can then calculate the probability $\mathbb{P}_\theta[X \leq x] = F(x; \theta)$ for any possible outcome x. Statistics deals with the inverse problem: suppose that we know the precise functional form of $F(x; \theta)$, but do not know which is the true θ. If we have an outcome x (a realisation of X), is it possible to say something useful about θ? It seems that we should be able to do so. Since θ influences what outcomes are most probable, then knowing an outcome should give us information on which θ are plausible. The topic of this text will be how exactly to make this connection rigorous and show how to exploit it in order to (a) make the best possible use of our data x to better inform ourselves about θ and (b) understand how certain we can be about our inferences on θ for the given data x. In summary, our framework is as follows:

1. There is a distribution $F(x; \theta)$ depending on an unknown $\theta \in \mathbb{R}^p$.
2. We observe the realisation of n independent identically distributed random variables $X_1, \ldots, X_n$ that follow this distribution.

3. We wish to use our n observations (the realisations of $X_1, \ldots, X_n$) in order to make statements about the true value of θ and to quantify the uncertainty associated with those statements.

At first glance, this framework may seem restrictive. Indeed, it represents a significant simplification over the much broader framework where one can develop statistical methodology. For example, in general, the unknown parameter of interest θ might not be an element of $\mathbb{R}^p$, but an element of a more general mathematical space (a space of functions, for instance). Also the data $(X_1, \ldots, X_n)$ could be dependent; they could themselves be vectors, or functions, or some more general mathematical object.

However, some of the key ideas that statisticians employ in order to attack these more general situations are already present in the simpler scenario that we will consider in this text. In fact, many highly more complex situations can often be reduced to this simpler case by a careful use of mathematics (for example, a real function can be identified with a vector in $\mathbb{R}^p$ when represented by its basis coefficients in some basis expansion, a dependent collection of random variables might in fact be approximately independent, and so on). In a sense, the framework we will consider here is the simplest non-trivial case that nevertheless contains the germs of generality.

Following is an overview of the contents of this text:

1. In Chap. 1, we will review the different types of probability models that we will construct statistical methods for. We will try to understand what situations they are suitable for, and what are some of their key properties. We will also try to find a unifying framework in which we can describe several of these models at once: instead of developing results separately for each model, we will try to give an abstract description of some key common characteristics that will be useful for obtaining general results. At the end of the chapter, we will consider the problem of how to choose a type of model, whether by first principles or by means of exploratory data analysis using numerical and graphical summaries.

2. In Chap. 2, we will develop the relevant concepts and probabilistic results that are needed in order to study the problem of sampling from probability models. We will probe the behaviour of the random sample, and how this relates to the original model, and what aspects of a sample are important for the purposes of statistical inference. An important focus will be to describe the probabilistic behaviour of functions of a sample. That is, given a sample $X_1, \ldots, X_n$ from a distribution F, what is the distribution of $g(X_1, \ldots, X_n)$ for some function g? The reason we will do this is simple: all that we have available to do statistics is the sample, so anything we do will be a function of the sample!

3. Once we know what probability models we wish to consider, and how to handle samples from probability models, we will turn to the most basic statistical inference question one can ask: given a sample $X_1, \ldots, X_n$ from a distribution F_θ that depends on an unknown parameter θ, construct an estimator: a function of the sample whose purpose is to estimate θ. We will consider how to formalise

the quality of such an estimator in terms of quantifying its accuracy, and what are methods for constructing good estimators (for example, are there optimal methods?).

4. Chapter 4 deals with a somewhat different problem. Instead of trying to guess which θ was the one that generated the observed sample $X_1, \ldots, X_n$, we will attempt to answer the following question: given a candidate value θ_0 for θ (or some candidate values forming a set Θ_0), decide on the basis of the sample $X_1, \ldots, X_n$ whether this value (or set of values) is good guess for the true θ. An important part of the chapter will be devoted to making formal what we mean by candidate values, good guesses (and bad guesses), and whether there are optimal strategies to do so. We will also be considering how to quantify the accuracy of our decisions.

5. Finally, in Chap. 5, we will deal with the third of the basic trio of problems of statistical inference: confidence intervals. Roughly speaking, instead of trying to estimate the precise value of θ that generated our sample $X_1, \ldots, X_n$, we wish to provide a whole range of values in the form of some interval, which will very likely contain the true parameter θ. This chapter will formalise this notion and consider how we can construct "small" regions that have high probability of covering the true parameter θ. We will, in fact, see that the problem of constructing confidence intervals is very closely connected both with the problem of point estimation and with the problem of hypothesis testing.

Regular Probability Models

1

Before setting out to explore how we can use statistics in order to learn about the structure of probability models given data from these models, we must first specify what types of probability models we shall consider (and some of their basic properties). For the purposes of this course, a probability model will be the distribution F of a random variable X which takes values in some subset of the real line $\mathbb{R}$:

$$F(x) = \mathbb{P}[X \leq x], \qquad x \in \mathbb{R}.$$

We write $X \sim F$ to state that F is the distribution of X. If $\{X_i\}_{i \in I}$ is a collection of independent identically distributed random variables with distribution F, we write $X_i \overset{\text{iid}}{\sim} F$. The distribution F will typically depend on one or several parameters that we shall represent as $\theta = (\theta_1, \ldots, \theta_p)^\top \in \Theta \subseteq \mathbb{R}^p$ (depending on the context, a different Greek letter or a Latin letter may be used). The space Θ where the parameter θ belongs is called the *parameter space*. To indicate that the distribution F depends on the parameter θ, we will often write F_θ or $F(x; \theta)$. All of the examples we will see and most of the theory we will develop will pertain to probability models that we shall call *regular*.

Definition 1.1 (Regular Parametric Probability Models)

Let X be a real-valued random variable, and let F_θ be its distribution function, for θ a parameter with parameter space $\Theta \subseteq \mathbb{R}^p$. The probability model $\{F_\theta : \theta \in \Theta\}$ will be called regular if one of the two following conditions holds:
1. For all $\theta \in \Theta$, the distribution F_θ is continuous with density $f(x; \theta)$.
2. For all $\theta \in \Theta$, the distribution F_θ is discrete with probability mass function $f(x; \theta)$ such that $\sum_{x \in \mathbb{Z}} f(x; \theta) = 1$ for all $\theta \in \Theta$.

Simply put, the model F_θ cannot switch between continuous and discrete depending on the value of θ. And, if it is discrete, the sample will always be taken to be a subset of the integers (e.g. it cannot be $\mathbb{Z} + \theta$, where $\theta \in [0, 1]$). The set

© Springer International Publishing Switzerland 2016
V.M. Panaretos, *Statistics for Mathematicians*, Compact Textbooks in Mathematics,
DOI 10.1007/978-3-319-28341-8_1

$\mathcal{X} := \{x \in \mathbb{R} : f(x; \theta) > 0\}$ will be called the *sample space* of X (note that $\mathcal{X}$ could depend on θ, but it will always satisfy $\mathcal{X} \subseteq \mathbb{R}$ in the continuous case, or $\mathcal{X} \subseteq \mathbb{Z}$ in the discrete case).

We will now review several regular probability models and their basic characteristics, explain what situations they are appropriate as models for, and give some illustrative examples.

▶ **Remark 1.2 (Notation $\mathbb{P}_\theta$ and $\mathbb{E}_\theta$)** When F depends on a parameter θ, we still have

$$F(x; \theta) = \mathbb{P}[X \leq x].$$

Since the left-hand side depends on θ, the right-hand side also must depend on θ, even though this is not explicit in our notation. Sometimes we will need to make that clear, in which case we will write $\mathbb{P}_\theta$ instead of just $\mathbb{P}$ in order to remind ourselves of this dependence. Similarly, we will sometimes write $\mathbb{E}_\theta$ instead of just $\mathbb{E}$ for the expectation of X when its distribution is $F(x; \theta)$.

1.1   Discrete Regular Models

Perhaps the simplest imaginable probability model is the Bernoulli distribution. This models a situation where there are only two possible outcomes, often termed "success" and "failure". The prototypical example is that of flipping a coin, where success (say heads) has probability p and failure (tails) has probability $1 - p$.

Definition 1.3 (Bernoulli Distribution)

A random variable X is said to follow the Bernoulli distribution with parameter $p \in (0, 1)$, denoted $X \sim \text{Bern}(p)$, if
1. $\mathcal{X} = \{0, 1\}$,
2. $f(x; p) = p\mathbf{1}\{x = 1\} + (1 - p)\mathbf{1}\{x = 0\}$.
   The mean, variance and moment generating function of $X \sim \text{Bern}(p)$ are given by

$$\mathbb{E}[X] = p, \qquad \text{Var}[X] = p(1 - p), \qquad M(t) = 1 - p + pe^t.$$

Example 1.4

Almost any random phenomenon whose outcomes may be classified in one of two categories can be modelled via the Bernoulli distribution. We simply name one category as success and the other as failure (success is usually the case we are most interested in).
1. Sample a voter from some large electorate (so large that we take it to be countably infinite) right after the ballots have closed, and let X be the vote she cast in the referendum. Then $X = 1$ (yes) with probability p and $X = 0$ (no) with probability $1 - p$, where p is the proportion of voters who voted yes.

2. Consider a sonogram that is made with the purpose of determining the sex of a foetus. The outcome X can either be $X = 1$ (girl) or $X = 0$ (boy), with some probabilities p and $1 - p$, respectively. The value of p in this case is determined by many and diverse environmental factors, but in general can be considered to be constant within homogeneous populations.
3. Consider a quantum measurement on the spin of an electron in a particle system. The outcome can either be 1 (spin up) or 0 (spin down) with probabilities p and $1 - p$. The value of the parameter here depends on the particular physical properties of the system.
4. Consider the barometric pressure in the lake Geneva region on a typical summer day. This might be high (if above a certain threshold) or low (otherwise), and these two outcomes may be coded as 1 and 0, respectively. Their corresponding probabilities, p and $1 - p$, are determined by several environmental factors.
5. More generally, we may create a Bernoulli random variable Y from any other random variable X in the following way. Let $A \subseteq \mathcal{X}$ be some event in the sample space of X, and define $Y = \mathbf{1}\{X \in A\}$. Then Y has a Bernoulli distribution with $p = \mathbb{P}[X \in A]$. Here, we interpret success as the realisation of X lying in A.

$\square$

More often than not, we have several independent repetitions of an experiment with two possible outcomes, say "success" and "failure" and we wish to model the total number of successes. If the individual experiments are modelled as Bernoulli experiments, then we are inevitably led to the *binomial distribution*. This models the total number of heads in a sequence of n independent coin flips.

Definition 1.5 (Binomial Distribution)

A random variable X is said to follow the binomial distribution with parameters $p \in (0, 1)$ and $n \in \mathbb{N}$, denoted $X \sim \text{Binom}(n, p)$, if
1. $\mathcal{X} = \{0, 1, 2, \ldots, n\}$,
2. $f(x; n, p) = \binom{n}{x} p^x (1 - p)^{n-x}$.

The mean, variance and moment generating function of $X \sim \text{Binom}(n, p)$ are given by

$$\mathbb{E}[X] = np, \qquad \text{Var}[X] = np(1 - p), \qquad M(t) = (1 - p + pe^t)^n.$$

Exercise 1 Show that if $X = \sum_{i=1}^{n} Y_i$ where $Y_i \overset{iid}{\sim} \text{Bern}(p)$, then $X \sim \text{Binom}(n, p)$.

Example 1.6

Since the binomial is a sum of independent Bernoulli random variables, we can expect that our previous examples can be extended to give us examples of using the binomial distribution (though this is not the case with all of them: we need both independence *and* equal probabilities of success for a binomial distribution to be induced).
1. Sample n voters from the same infinite electorate right after the ballots have closed, and let Y be the number of voters in that sample who voted "yes". Then Y is binomial with n trials and success probability p, where p is the proportion of voters who voted yes.

2. Consider a particle system with the property that the spin of individual particles is independent of all others. If there are n particles, then the number Y of spin up particles is binomially distributed with parameters n and p, where p is as before, and is related to the electromagnetic properties of the system.
3. Consider again the barometric pressure in the lake Geneva region on a typical summer day, which can be high or low, with corresponding probabilities p and $1 - p$. Let Y be the number of days with high barometric pressure within a period of n consecutive days. Though Y is a sum of Bernoulli variables, it is not a Binom(n, p). The reason is that the pressure conditions are dependent between consecutive days (hence the Bernoulli trials are not independent).
4. Going back to the sonogram example, suppose that the probability of a given foetus being of female sex is p. Consider now a sonogram whose purpose is that of of determining the number of foetuses of female sex among two foetuses being gestated by the same woman (twins). The outcome Y can either be 0, or 1 or 2. If we know whether the twins are non-identical (say this is an event called A), then:

$$\mathbb{P}[Y = 0|A] = (1 - p)^2, \mathbb{P}[Y = 1|A] = 2p(1 - p), \mathbb{P}[Y = 2|A] = p^2.$$

In other words, given that the twins are non-identical

$$\mathbb{P}[Y = y|A] = \binom{2}{y} p^y (1 - p)^{2-y}, \qquad y = 0, 1, 2,$$

and so Y is indeed binomial given A. However, if we do not know whether the twins are non-identical, we factor in the possibility that the twins might be identical. In this case:

$$\mathbb{P}[Y = y] = \mathbb{P}[Y = y|A]\mathbb{P}[A] + \mathbb{P}[Y = y|A^c]\mathbb{P}[A^c]$$

$$= \binom{2}{y} p^y (1 - p)^{2-y}\mathbb{P}[A] + \left(p\mathbf{1}\{y = 2\} + (1 - p)\mathbf{1}\{y = 0\}\right)\mathbb{P}[A^c].$$

If $\mathbb{P}[A^c] \neq 0$, this expression will in general not be expressible as a binomial probability mass function, and so Y may not be binomial. This example highlights that dependence between trials may be subtly disguised, and that one must think carefully about the nature of the probability experiment before proceeding with a specific model. $\square$

Suppose now that we start a sequence of independent Bernoulli trials, say coin flips, and we continue flipping the coin until the first time we get heads (success). The number of tails (failures) until the first apparition of heads (the first success) has the *geometric distribution*.

Definition 1.7 (Geometric Distribution)

A random variable X is said to follow the Geometric distribution with parameter $p \in (0, 1)$, denoted $X \sim \text{Geom}(p)$, if
1. $\mathcal{X} = \{0\} \cup \mathbb{N}$,
2. $f(x; p) = (1 - p)^x p$.

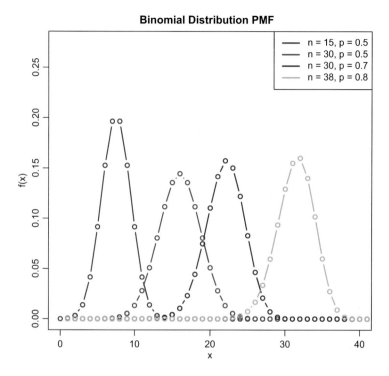

Fig. 1.1 Binomial probability mass functions for different values of the parameters n and p

The mean, variance and moment generating function of $X \sim \text{Geom}(p)$ are given by

$$\mathbb{E}[X] = \frac{1-p}{p}, \quad \text{Var}[X] = \frac{(1-p)}{p^2}, \quad M(t) = \frac{p}{1-(1-p)e^t}, \quad \text{for } t < -\log(1-p).$$

Exercise 2 Let $\{Y_i\}_{i \geq 1}$ be an infinite collection of random variables, where $Y_i \overset{\text{iid}}{\sim}$ Bern(p). Let $T = \min\{k \in \mathbb{N} : Y_k = 1\} - 1$. Then $T \sim \text{Geom}(p)$ (Figs. 1.1 and 1.2).

What about the distribution of the number of failures until the rth success in a sequence of Bernoulli trials? This follows the *negative binomial distribution* (also known as the *Pólya distribution*).

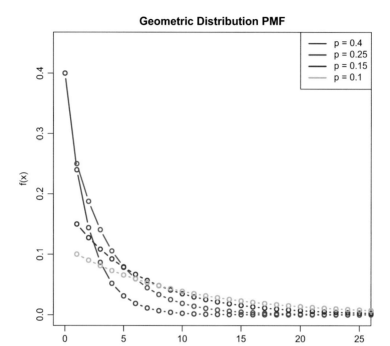

Fig. 1.2 Geometric probability mass functions for different values of the parameter p

---

Definition 1.8 (Negative Binomial Distribution)

A random variable X is said to follow the negative binomial distribution with parameters $p \in (0, 1)$ and $r > 0$, denoted $X \sim \text{NegBin}(r, p)$, if

1. $\mathcal{X} = \{0\} \cup \mathbb{N}$,

2. $f(x; p, r) = \binom{x + r - 1}{x} (1 - p)^x p^r$.

The mean, variance and moment generating function of $X \sim \text{NegBin}(r, p)$ are given by

$$\mathbb{E}[X] = r \frac{1 - p}{p}, \quad \text{Var}[X] = r \frac{(1 - p)}{p^2}, \quad M(t) = \frac{p^r}{[1 - (1 - p)e^t]^r}, \quad \text{for } t < -\log(1 - p).$$

Exercise 3 Show that if $X = \sum_{i=1}^{r} Y_i$ where $Y_i \overset{\text{iid}}{\sim} \text{Geom}(p)$, then $X \sim \text{NegBin}(r, p)$. Deduce the mean, variance and moment generating function of X.

What if we would like to count the number of successes not within a discrete set of trials but within a bounded uncountably infinite set, such as an interval? For example, the total number of calls in a call centre within a 10-min interval. In principle, the phone could ring at any instant of time—but there are uncountably infinite instants (=trials) within the 10-minute interval! It turns out that such a distribution exists, provided that the probability of a success for any given instant is "very small", and it is called the *Poisson distribution*.

Definition 1.9 (Poisson Distribution)

A random variable X is said to follow the Poisson distribution with parameter $\lambda > 0$, denoted $X \sim \text{Poisson}(\lambda)$, if
1. $\mathcal{X} = \{0\} \cup \mathbb{N}$,
2. $f(x; \lambda) = e^{-\lambda} \dfrac{\lambda^x}{x!}$.

The mean, variance and moment generating function of $X \sim \text{Poisson}(\lambda)$ are given by

$$\mathbb{E}[X] = \lambda, \qquad \text{Var}[X] = \lambda, \qquad M(t) = \exp\{\lambda(e^t - 1)\}.$$

Exercise 4 Let $X_i \overset{iid}{\sim} \text{Poisson}(\lambda)$. Show that $Y = \sum_{i=1}^n X_i \sim \text{Poisson}(n\lambda)$.

Exercise 5 Let $X \sim \text{Poisson}(\lambda)$ and $Y \sim \text{Poisson}(\mu)$ be independent. Show that the conditional distribution of X given $X + Y = k$ is $\text{Binom}(k, \lambda/(\lambda + \mu))$.

It would seem that the Poisson distribution came out of nowhere, whereas the other distributions we considered were linked with the Bernoulli distribution. It turns out that there is an important connection between the Poisson and Binomial distributions. Roughly speaking, a Poisson distribution is the limit of a Binomial distribution when $n \to \infty$ and $p = \lambda/n$ (the number of trials diverges to infinity but the probability of success decreases to zero linearly with respect to the number of trials). This link also helps us make precise mathematical sense of the way we motivated the Poisson distribution. It is the *Law of Rare Events*, and will be stated rigorously in Exercise 24 (p. 54) (Figs. 1.3 and 1.4).

Example 1.10

We list here some random experiments for which the Poisson distribution is a reasonable probability model. All of these involve modelling counts over a finite time horizon, when there is no a priori upper bound on the total.
1. The number of visits to a website during a given day can be well modelled by a Poisson distribution. The parameter of the Poisson distribution will be interpreted as the mean number of visits on that day.
2. The yearly number of earthquakes in a given bounded spatial region is typically Poisson distributed, with parameter equal to the mean number of earthquakes per year in that region.

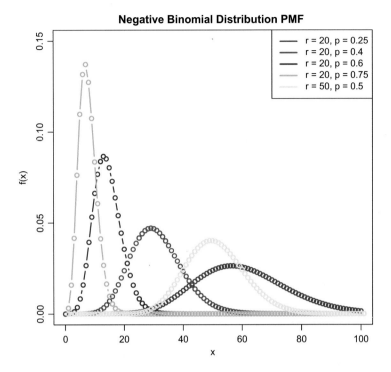

Fig. 1.3 Negative binomial probability mass functions for different values of r and p

3. Radioactive materials have unstable atoms, which emit particles (such as alpha particles and gamma rays). Quantum theory postulates that, at the level of each atom, the number of particles emitted within a given fixed time interval is random. The typical model for this random variable is a Poisson distribution with mean given by the decay constant of the material.

4. In positron emission tomography, we attempt to image the interior of the human body in order to detect features of interest, for example cancers. A tracer is injected into the human body that emits positrons. This tracer is spread throughout the human body, but concentrates more in tissue with high metabolic activity (e.g. a cancerous tissue). By counting the number of positrons emitted at a given physical location, we have an indication of the metabolic activity in that location. The number of particles emitted at a given location typically behaves like a Poisson distribution with mean parameter given by the concentration of the tracer at that physical location. In other words, the intensity of the tomography image obtained at any pixel is Poisson distributed with mean given by the true concentration of the material at that pixel.

□

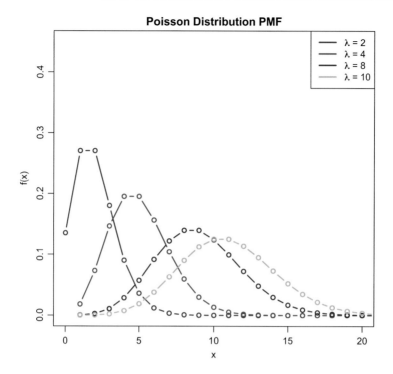

Fig. 1.4 Poisson probability mass functions for different values of the parameter λ

1.2   Continuous Regular Models

We now switch to the continuous case, and consider some of the key probability models for random variables taking values in $\mathbb{R}$. To define these, it suffices to determine their probability density function. We first consider one of the simplest continuous probability models: a random variable that is "equally likely" to take values anywhere on a bounded interval.

Definition 1.11 (Uniform Distribution)

A random variable X is said to follow the uniform distribution with parameters $-\infty < \theta_1 < \theta_2 < \infty$, denoted $X \sim \text{Unif}(\theta_1, \theta_2)$, if

$$f_X(x; \theta) = \begin{cases} (\theta_2 - \theta_1)^{-1} & \text{if } x \in (\theta_1, \theta_2), \\ 0 & \text{otherwise.} \end{cases}$$

The mean, variance and moment generating function of $X \sim \text{Unif}(\theta_1, \theta_2)$ are given by

$$\mathbb{E}[X] = (\theta_1 + \theta_2)/2, \quad \text{Var}[X] = (\theta_2 - \theta_1)^2/12, \quad M(t) = \frac{e^{t\theta_2} - e^{t\theta_1}}{t(\theta_2 - \theta_1)}, \quad t \neq 0, M(0) = 1.$$

In a discrete setting, the uniform distribution gives equal probability to any possible simple outcome from within the finite sample space of outcomes. In the continuous case, the probability of observing a specific number in (θ_1, θ_2) is precisely zero, but uniformity is understood in the sense that the probability of observing an outcome falling in a given subinterval of (θ_1, θ_2) is proportional to the length of that interval.

Example 1.12

The uniform distribution is as spread out as possible over a finite interval. In that sense, it can be used to model situations where we have "complete ignorance", where we are not prepared to make any assumptions, or where the phenomenon under study is highly unpredictable.

1. Suppose that our bus is supposed to pass every 10 min, and we arrive at a random moment at the bus stop, without knowing the schedule. It is natural to model our waiting time by a uniform distribution on $(0, 10)$.
2. Suppose that our compass is broken, and the needle moves freely. Then, if we move in the direction that the compass indicates for "north" at some random moment, the true direction we will move in can be modelled as a random variable with the uniform distribution on $(0, 2\pi)$ (where we can imagine $\pi/2$ to correspond to the true "north").
3. Consider the movement of excited gas molecules (in high temperature) in a container shaped as a cube of edge length 1. If we let the molecules move freely inside the container, and then ask for the location of a specific molecule after some time t (where t is large), the coordinates of this location (X, Y, Z) can be modelled very accurately by iid uniform random variables on $(0, 1)$, regardless of the starting point of the molecule.

□

Our next model is typically appropriate when we wish to model the time elapsed until the occurrence of a certain event, or between events, when this time is random.

Definition 1.13 (Exponential Distribution)

A random variable X is said to follow the exponential distribution with parameter $\lambda > 0$, denoted $X \sim \text{Exp}(\lambda)$, if

$$f_X(x; \lambda) = \begin{cases} \lambda e^{-\lambda x}, & \text{if } x \geq 0 \\ 0 & \text{if } x < 0. \end{cases}$$

The mean, variance and moment generating function of $X \sim \text{Exp}(\lambda)$ are given by

$$\mathbb{E}[X] = \lambda^{-1}, \qquad \text{Var}[X] = \lambda^{-2}, \qquad M(t) = \frac{\lambda}{\lambda - t}, \quad t < \lambda.$$

Note the interpretation here: λ^{-1} is the average time until the occurrence of the event of interest (measured in some given unit of time). So λ is interpreted as a rate parameter. The exponential distribution can be considered to be the continuous version of the geometric distribution, when the number of trials becomes large, and the probability of success becomes small.

A crucial property of the exponential distribution is that it is "memoryless": no matter how long you've been waiting already, the probability of waiting for an additional amount of time x only depends on x, and not your past waiting time:

Exercise 6 Let $X \sim \text{Exp}(\lambda)$. Then $\mathbb{P}[X \geq x + t \,|\, X \geq t] = \mathbb{P}[X \geq x]$.

The exponential distribution is, in fact, the unique distribution on $[0, \infty)$ with this property (see Exercise 14, p. 27). Therefore, when choosing the exponential distribution as a model for a random time, we must always ask if it is reasonable to assume that this random time has the lack of memory property (Figs. 1.5 and 1.6).

Example 1.14

The exponential distribution has important connections to the Poisson distribution. Roughly speaking, if the time between consecutive occurrences of a certain phenomenon is independent

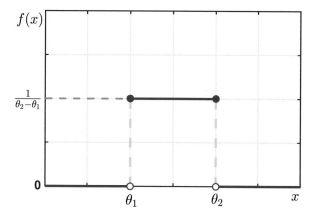

Fig. 1.5 Uniform probability density function for general values of (θ_1, θ_2)

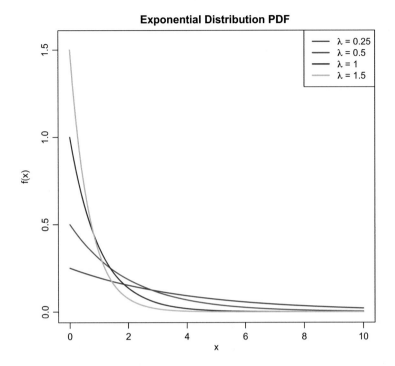

Fig. 1.6 Exponential probability density functions for different values of the parameter λ

exponential, then the number of occurrences of the phenomenon up to a given time will be Poisson. For example:

1. The time between two consecutive occurrences of an earthquake at a given spatial region can be modelled as an exponential random variable.
2. The time between consecutive emissions of alpha particles from an atom of a radioactive material is very well modelled by an exponential distribution. The rate of this exponential distribution will be intimately related to the decay constant of the material.
3. The amount of time between two consecutive visits at a website can also be modelled by an exponential distribution.

$\square$

Exercise 7 Let X, Y be independent exponential random variables with rates λ_1 and λ_2. Prove that $Z = \min\{X, Y\}$ is also exponential with rate $\lambda_1 + \lambda_2$.

Now suppose that we are interested in the time until the rth event, in a situation where the times between events are distributed as iid $\text{Exp}(\lambda)$. This resembles the discrete situation where we are waiting until the rth success in a sequence of Bernoulli trials, which takes us from the geometric to the negative binomial (the negative binomial distribution being the sum of r iid geometric random variables).

It turns out that the sum of r iid exponential random variables has a gamma distribution:

Definition 1.15 (Gamma Distribution)

A random variable X is said to follow the gamma distribution with parameters $r > 0$ and $\lambda > 0$ (the *shape* and *scale* parameters, respectively), denoted $X \sim$ Gamma(r, λ), if

$$f_X(x; r, \lambda) = \begin{cases} \frac{\lambda^r}{\Gamma(r)} x^{r-1} e^{-\lambda x}, & \text{if } x \geq 0 \\ 0 & \text{if } x < 0. \end{cases}$$

The mean, variance and moment generating function of $X \sim$ Gamma(r, λ) are given by

$$\mathbb{E}[X] = r/\lambda, \qquad \text{Var}[X] = r/\lambda^2, \qquad M(t) = \left(\frac{\lambda}{\lambda - t}\right)^r, \quad t < \lambda.$$

Note that the way we have defined the gamma distribution does not restrict r to be a natural number. It is indeed true that a gamma distribution can be defined more generally for $r > 0$. The interpretation as a sum of r exponentials of rate λ will only be valid when r happens to be a positive integer. The gamma distribution can provide a flexible model for a wide variety of phenomena that give rise to non-negative random variables. The suitability of these models is not always completely founded on concrete physical principles. It is sometimes dictated by convenience, and other times by extensive practical experience.

The function $\Gamma(y)$ is the gamma function (from which the distribution inherits its name). In the special case when r is a positive integer, $\Gamma(r) = (r - 1)!$. There is a particular special case of the Gamma distribution, known as the chi-squared distribution, that is especially important in statistical theory and practice:

Definition 1.16 (Chi-Square Distribution)

A random variable X is said to follow the chi-square distribution with parameter $k \in \mathbb{N}$ (called the number of degrees of freedom), denoted $X \sim \chi_k^2$, if it holds that $X \sim$ Gamma$(k/2, 1/2)$. In other words,

$$f_X(x; k) = \begin{cases} \frac{1}{2^{k/2} \Gamma(\frac{k}{2})} x^{\frac{k}{2}-1} e^{-\frac{x}{2}}, & \text{if } x \geq 0 \\ 0 & \text{if } x < 0. \end{cases}$$

The mean, variance and moment generating function of $X \sim \chi_k^2$ are given by

$$\mathbb{E}[X] = k, \qquad \text{Var}[X] = 2k, \qquad M(t) = (1 - 2t)^{-k/2}, \quad t < \frac{1}{2}.$$

Exercise 8 Show that $X \sim \chi_2^2$ if and only if $X \sim \text{Exp}(1/2)$.

The continuous probability models we have encountered so far have all been restricted either to a bounded interval or to the positive reals. In many phenomena, we expect that the random variable can assume any positive value, but its distribution is centred at (and is symmetric about) a centre location μ. The parameter μ represents the "location" or the value around which we expect typical realisations of the random variable to lie. Further to the location, there is typically a "scale" parameter, say σ^2, which expresses how concentrated or diffuse the distribution is around the centre. A broad such family of models is the so-called *location-scale* family of models. Among location-scale models, the most important and well-studied, and perhaps the most widely applicable is the *normal distribution*, also referred to as the *Gaussian distribution*.

Definition 1.17 (Normal Distribution)

A random variable X is said to follow the normal distribution with parameters $\mu \in \mathbb{R}$ and $\sigma^2 > 0$ (the *mean* and *variance* parameters, respectively), denoted $X \sim \text{N}(\mu, \sigma^2)$, if

$$f_X(x; \mu, \sigma^2) = \frac{1}{\sigma\sqrt{2\pi}} \exp\left\{ -\frac{1}{2}\left(\frac{x-\mu}{\sigma}\right)^2 \right\}, \quad x \in \mathbb{R}.$$

The mean, variance and moment generating function of $X \sim \text{N}(\mu, \sigma^2)$ are given by

$$\mathbb{E}[X] = \mu, \qquad \text{Var}[X] = \sigma^2, \qquad M(t) = \exp\{t\mu + t^2\sigma^2/2\}.$$

▶ **Remark 1.18** In the special case $Z \sim N(0, 1)$, we use the notation $\varphi(z) = f_Z(z)$ and $\Phi(z) = F_Z(z)$, and call these the *standard normal density* and *standard normal CDF*, respectively.

Example 1.19

The normal distribution can be a very good model for a bewildering variety of phenomena. Intuitively, almost any phenomenon that can be thought to arise as the result of the addition of a large number of random variables with finite variances can be modelled via a normal distribution (see the Central Limit Theorem for a precise statement, Theorem 2.23 (p. 56)). In general, the normal distribution will be a good model for random variables with finite variance, whose distribution is symmetric about a certain value μ, and whose probability of being far from μ decays fast.

1. Measurement error is most typically modelled as a normal random variable. Suppose that we are trying to measure a quantity μ, and our measurement device is imperfect, thus yielding measurements Y corrupted by error ε. If the error is additive, then a natural probability model is to assert that $Y = \mu + \varepsilon$, and $\varepsilon \sim N(0, \sigma^2)$. Consequently, $Y \sim N(\mu, \sigma^2)$.
2. It is well established that several random physical phenomena are distributed according to the normal distribution. For example, the position after time t of a molecule that moves on

a line subject to collisions from other molecules has a normal distribution with a mean at its starting point and variance equal to t. The velocity of any particle in a one-dimensional space under thermodynamic equilibrium will be normally distributed. The ground state of a quantum harmonic oscillator will also be normally distributed.

3. The re-scaled difference between a random variable and its mean can very often be approximated by a normal distribution. Typically this depends on taking a limiting argument over some parameter of that random variable. This includes variables that are discrete. For example, we will see later that the approximation is valid in the case of a binomial distribution with a large number of trials, or a Poisson distribution with a large rate parameter (in both cases, after appropriate centering and scaling).

4. Experience shows that a wide range of phenomena in the biological sciences, when suitably transformed, are remarkably well approximated by the normal distribution. The same is true of phenomena in the social sciences, economics and finance. In most of these cases, the underlying effect is a central limit theorem effect (Figs. 1.7 and 1.8).

□

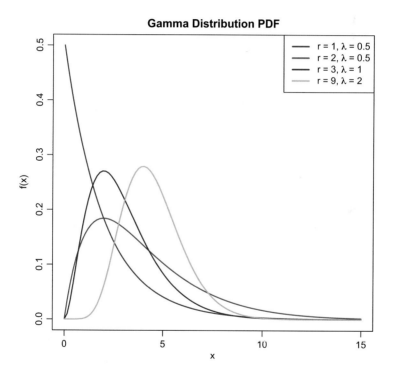

Fig. 1.7 Gamma probability density functions for different values of the parameters r and λ

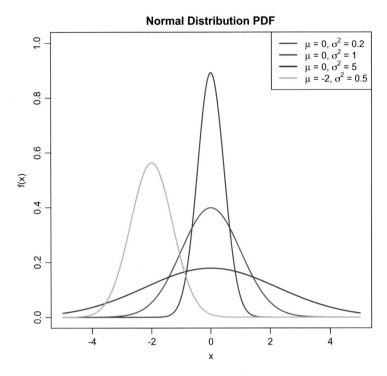

Fig. 1.8 Normal probability density functions for different values of the parameters μ and σ^2

1.3   Exponential Families of Distributions

Though it may not be immediately obvious at first sight, many of the models we considered earlier—whether discrete or continuous—have some important similarities in terms of their structure and their properties. For this reason, we will introduce in this paragraph an additional level of abstraction, and consider most of the previous models as special cases of a broader family of probability models called the *exponential family of distributions*. The advantage of such an approach is that, once we have this more abstract definition, any properties we prove for the general case will immediately be inherited by all the special cases. Here is the definition:

Definition 1.20 (The Exponential Family of Distributions)

A regular probability distribution is said to be a member of a k-parameter exponential family, if its density (or frequency) admits the representation

$$f(x) = \exp\left\{\sum_{i=1}^{k} \phi_i\, T_i(x) - \gamma(\phi_1, \ldots, \phi_k) + S(x)\right\}, \qquad x \in \mathcal{X}, \qquad (1.1)$$

where:
1. $\phi = (\phi_1, \ldots, \phi_k)$ is a k-dimensional parameter in $\mathbb{R}^k$;
2. $T_i : \mathcal{X} \to \mathbb{R}, i = 1, \ldots, k, S(x) : \mathcal{X} \to \mathbb{R}$, and $\gamma : \mathbb{R}^k \to \mathbb{R}$ are real-valued functions;
3. The sample space $\mathcal{X}$ does not depend on ϕ.

▶ **Remark 1.21** The parameter ϕ is called the natural parameter.

▶ **Remark 1.22** The fact that there is an exponential in the formula (1.1) is in itself not the most important structural property of an exponential family (since any density function can be written as $f(x) = \exp\{\log f(x)\}$ on its support). The important property is that the density can be factorised into three parts: one that only depends on ϕ, i.e. $\exp\{-\gamma(\phi)\}$; one that only depends on x, i.e. $\exp\{S(x)\}$; and one that depends on both ϕ and x but in a very special way: as a linear combination of the coordinates of ϕ with coefficients that are functions of x.

▶ **Remark 1.23** The exponential family of distributions should not be confused with the exponential distribution. It is unfortunate that they share such a similar name. To avoid confusion, we will always speak of an exponential **family** to distinguish from an exponential **distribution**.

We will see that all the distributions that we have so far seen, except for the uniform distribution, constitute exponential families. In order to see this, we will need to manipulate the expressions of the corresponding densities (or frequencies) in order to bring them to the form given by the form given by Eq. (1.1). It will often happen that the usual parameter employed does not coincide with the natural parameter. However, it will typically be the case that $\phi = \eta(\theta)$ for some twice differentiable 1–1 mapping $\eta : \Theta \to \mathbb{R}^k$ (and so $\gamma(\phi) = \gamma(\eta(\theta)) = d(\theta)$, for $d = \gamma \circ \eta$). In this form, the exponential family density/frequency will take the form:

$$\exp\left\{\sum_{i=1}^{k} \phi_i T_i(x) - \gamma(\phi) + S(x)\right\} = \exp\left\{\sum_{i=1}^{k} \eta_i(\theta) T_i(x) - d(\theta) + S(x)\right\}.$$

Either formulation can be used, depending on which is most convenient in a specific context: for the purpose of doing theory and proving general results, the *natural representation* (also called *natural parametrisation*) given by $\exp\left\{\sum_{i=1}^{k} \phi_i T_i(x) - \gamma(\phi) + S(x)\right\}$ is more convenient.[1] In most practical settings,

---

[1] The reason for this is that in the natural representation, the parameter appears linearly in the exponent. In the usual representation, the parameter appears nonlinearly, as the image through the function η. This complicates things when we will need to differentiate with respect to the parameter.

problems are presented in such a way that the parameter of interest is the θ parameter from the *usual representation* (also called *usual parametrisation*) given by $\exp\left\{\sum_{i=1}^{k} \eta_i(\theta) T_i(x) - d(\theta) + S(x)\right\}$. Generally, thus, the strategy is to prove any necessary theorems in the natural representation, and then translate them into results for the usual representation.

Example 1.24 (Binomial Exponential Family)

Let $X \sim \text{Binom}(n, p)$. Recall that this means that $\mathcal{X} = \{0, 1, 2, \ldots, n\}$ and $f(x; p) = \binom{n}{x} p^x (1-p)^{n-x}$. Now, we may take the log and then exponentiate to obtain:

$$\binom{n}{x} p^x (1-p)^{n-x} = \exp\left\{\log\left(\frac{p}{1-p}\right) x + n \log(1-p) + \log\binom{n}{x}\right\}.$$

Define:

$$\phi = \log\left(\frac{p}{1-p}\right), \quad T(x) = x, \quad S(x) = \log\binom{n}{x}, \quad \gamma(\phi) = n\log(1+e^\phi) = -n\log(1-p).$$

Thus, if n is held fixed and only p is allowed to vary, the support of f does not depend on ϕ and so we see that the Binomial with fixed n is a 1-parameter exponential family. Here the usual parameter p is a twice differentiable bijection of the natural parameter ϕ:

$$p = \frac{e^\phi}{1 + e^\phi} \quad \& \quad \phi = \eta(p) = \log\left(\frac{p}{1-p}\right).$$

Here $p \in (0, 1)$ but $\phi \in \mathbb{R}$. $\qquad\qquad\qquad\qquad\qquad\qquad\qquad\qquad\qquad\qquad\quad\square$

Example 1.25 (Counterexample: Uniform Distribution)

Let $X \sim \text{Unif}(\theta_1, \theta_2)$. Notice that $f(x; \theta_1, \theta_2)$ is positive if and only if $x \in [\theta_1, \theta_2]$. Therefore the support of f depends on the parameter, and thus the uniform distribution is not an exponential family. Notice, though, that if we fix θ_1 and θ_2 and consider the specific fixed density (rather than a whole family as θ_1 and θ_2 vary), then we do have an exponential family form, albeit a degenerate one with a single member. $\qquad\qquad\qquad\qquad\qquad\qquad\qquad\square$

Example 1.26 (Gaussian Exponential Family)

Let $X \sim N(\mu, \sigma^2)$. Then we may write:

$$f(x; \mu, \sigma^2) = \frac{1}{\sigma\sqrt{2\pi}} \exp\left\{-\frac{1}{2}\left(\frac{x-\mu}{\sigma}\right)^2\right\}$$

$$= \exp\left\{-\frac{1}{2\sigma^2} x^2 + \frac{\mu}{\sigma^2} x - \frac{1}{2}\log(2\pi\sigma^2) - \frac{\mu^2}{2\sigma^2}\right\}.$$

Define:

$$\phi_1 = \frac{\mu}{\sigma^2}, \quad \phi_2 = -\frac{1}{2\sigma^2}, \quad T_1(x) = x, \quad T_2(x) = x^2, \quad S(x) = 0, \quad \gamma(\phi_1,\phi_2) = -\frac{\phi_1^2}{4\phi_2} + \frac{1}{2}\log\left(-\frac{\pi}{\phi_2}\right),$$

and also observe that the support of f is always $\mathbb{R}$, regardless of the parameter values. It follows that the $N(\mu,\sigma^2)$ distribution is a 2-parameter exponential family. □

Exercise 9 (More Exponential Families) Show that the following distributions constitute exponential families (perhaps when one of their parameters is held fixed):
1. The Poisson distribution.
2. The geometric distribution.
3. The negative binomial distribution.
4. The exponential distribution.
5. The gamma distribution.
6. The chi-square distribution.

There are several more probability models that form exponential families. Though we have not studied them here explicitly, it is worth mentioning them: the Pareto distribution, the Weibull distribution, the Laplace distribution, the chi-squared distribution, the lognormal distribution, the inverse Gaussian distribution, the inverse gamma distribution, the normal-gamma distribution and the beta distribution, among others.

Later, we will prove some key theorems on estimation and hypothesis testing for exponential families; and these results will then be valid for any specific exponential family.

1.4   Transforming Probability Models

It is often the case that we have a model for a particular random phenomenon whose outcome is described by a random variable X, but we are really interested in modelling some aspect of this phenomenon, say $g(X)$, where g is a known function.

Example 1.27

Suppose that R is a positive random variable denoting the radius of coverage of a wireless antenna. Assume that $R \sim \text{Unif}[a, b]$, for some $0 < a < b$. What is the distribution of the area of coverage, $A = \pi R^2$? □

The purpose of this section is to investigate what the distribution of $g(X)$ is, given knowledge of the distribution of X; in other words, how the distribution of a random variable X is transformed, when the random variable X is transformed.

In the discrete case, things are relatively straightforward (though they rarely give simple closed form expressions for the resulting distributions).

> **Lemma 1.28** *Let X be a discrete random variable, and $Y = g(X)$. Then, the sample space of Y is $\mathcal{Y} = g(\mathcal{X})$ and*
>
> $$F_Y(y) = \mathbb{P}[g(X) \leq y] = \sum_{x \in \mathcal{X}} f_X(x)\mathbf{1}\{g(x) \leq y\}, \qquad \forall y \in \mathcal{Y} \quad (1.2)$$
>
> $$f_Y(y) = \mathbb{P}[g(X) = y] = \sum_{x \in \mathcal{X}} f_X(x)\mathbf{1}\{g(x) = y\}, \qquad \forall y \in \mathcal{Y}. \quad (1.3)$$

Proof It suffices to observe that $\mathbb{P}[Y = y] = \sum_{x \in \mathcal{X}: g(x) = y} \mathbb{P}[X = x], \forall\, y \in \mathcal{Y}$.

$\square$

In the case where X is continuous, things are a bit more subtle to state and prove: the obtention of general formulas is not possible for non-bijective g. If g is not a bijection, the problem has to be attacked by direct methods that are specific to the setup:

Example 1.29 (Squared Standard Normal Has χ_1^2 Distribution)

Let $Z \sim N(0, 1)$. We would like to find the distribution of $Y = Z^2$. Note that $F_Y(y) = \mathbb{P}[Y \leq y] = 0$ if $y < 0$. For $y \geq 0$ we have

$$F_Y(y) = \mathbb{P}[Z^2 \leq y] = \mathbb{P}[|Z| \leq \sqrt{y}]$$
$$= \mathbb{P}[-\sqrt{y} \leq Z \leq \sqrt{y}] = \Phi(\sqrt{y}) - \Phi(-\sqrt{y}) = \Phi(\sqrt{y}) - (1 - \Phi(\sqrt{y}))$$
$$= 2\Phi(\sqrt{y}) - 1.$$

We can also find the density by differentiating:

$$f_Y(y) = 2\frac{d}{dy}\Phi(\sqrt{y}) = 2\frac{d}{d\sqrt{y}}\Phi(\sqrt{y})\frac{d}{dy}\sqrt{y}$$

$$= 2\phi(\sqrt{y})\frac{y^{-1/2}}{2} = 2\frac{1}{\sqrt{2\pi}}e^{-y/2}\frac{y^{-1/2}}{2}$$

$$= \frac{1}{\sqrt{2}\sqrt{\pi}}e^{-y/2}y^{-1/2} = \frac{1}{2^{1/2}\Gamma(1/2)}y^{1/2-1}e^{-y/2}.$$

Notice that the last expression is the density of the χ_1^2 distribution (see Definition 1.16, p. 13). We therefore have

$$Z \sim N(0, 1) \implies Z^2 \sim \chi_1^2.$$

(1.4)

□

On the other hand, if g is a monotone differentiable transformation, then we may derive general explicit (closed form) expressions for the distribution and density of $g(X)$.

Lemma 1.30 *Let X be a continuous random variable on $\mathcal{X} \subseteq \mathbb{R}$ and let $g : \mathcal{X} \rightarrow \mathbb{R}$ be monotone and differentiable, with derivative positive on $\mathcal{X}$. Let $Y = g(X)$. Then, the sample space of Y is $\mathcal{Y} = g(\mathcal{X})$ and*
- *if g is increasing, then $F_Y(y) = F_X(g^{-1}(y))$, $\forall y \in \mathcal{Y}$,*
- *if g is decreasing, then $F_Y(y) = 1 - F_X(g^{-1}(y))$, $\forall y \in \mathcal{Y}$.*
In either case, we will have

$$f_Y(y) = \left| \frac{\partial}{\partial y} g^{-1}(y) \right| f_X(g^{-1}(y)), \qquad \forall y \in \mathcal{Y}.$$

Proof Assume initially that g' is positive everywhere on $\mathcal{X}$ (g is monotone increasing). This means that $x \le y \iff g(x) \le g(y)$. Then, for $y \in \mathcal{Y}$,

$$F_Y(y) = \mathbb{P}[g(X) \le y] = \mathbb{P}[X \le g^{-1}(y)] = F_X(g^{-1}(y)).$$

Therefore,

$$f_Y(y) = \frac{\partial}{\partial y} F_Y(y) = \frac{\partial}{\partial y} F_X(g^{-1}(y)) = f_X(g^{-1}(y)) \frac{\partial}{\partial y} g^{-1}(y) = f_X(g^{-1}(y)) \left| \frac{\partial}{\partial y} g^{-1}(y) \right|,$$

with the last equality following from the fact that g' is everywhere positive. Now consider the case where g is monotone decreasing (and so g' is negative everywhere). This means that $x < y \iff g(x) > g(y)$. Then, for $y \in \mathcal{Y}$,

$$1 - F_Y(y) = \mathbb{P}[g(X) > y] = \mathbb{P}[X < g^{-1}(y)] = F_X(g^{-1}(y)) - \underbrace{\mathbb{P}[X = g^{-1}(y)]}_{=0}.$$

But $f_Y(y) = -\frac{\partial}{\partial y}(1 - F_Y(y))$. Therefore,

$$f_Y(y) = -\frac{\partial}{\partial y}(1 - F_Y(y)) = -f_X(g^{-1}(y)) \frac{\partial}{\partial y} g^{-1}(y) = f_X(g^{-1}(y)) \left| \frac{\partial}{\partial y} g^{-1}(y) \right|,$$

since $-g'$ is everywhere negative. This completes the proof.

□

Exercise 10 (Log-Normal Distribution) Let $X \sim N(\mu, \sigma^2)$, and show that the density of $Y = e^X$ is given by

$$f_Y(y) = \frac{1}{y\sigma\sqrt{2\pi}} \exp\left(\frac{-(\ln y - \mu)^2}{2\sigma^2}\right), \quad 0 < y < \infty.$$

The distribution of Y is called the log-normal distribution.

Exercise 11 (Random Number Generation) Let $Y \sim \text{Unif}(0,1)$ and let F be a distribution function. Prove that the distribution function of the random variable $X = F^{-1}(Y)$ is given precisely by F, where we define $F^{-1}(y) = \inf\{t \in \mathbb{R} : F(t) \geq y\}$ (see Definition A.6, p. 161). Observe that with this result, we can generate realisations from any distribution, provided that we can generate realisations from the uniform distribution.

An easy corollary to the last two lemmas combined is the following:

Corollary 1.31 (Affine Transformations) *Let X be a random variable and $Y = g(X)$. If $g(x) = ax + b$, $a \neq 0$, then*

$$\forall y \in \mathcal{Y}, \qquad F_Y(y) = \begin{cases} F_X\left(\frac{y-b}{a}\right) & a > 0, \\ 1 - F_X\left(\frac{y-b}{a}\right) + \mathbb{P}\left(X = \frac{y-b}{a}\right) & a < 0, \end{cases}$$

with $\mathbb{P}\left(X = \frac{y-b}{a}\right) = 0$ when X is a continuous random variable. Thus, for $y \in \mathcal{Y}$:

1. $f_Y(y) = |a^{-1}| f_X\left(\dfrac{y-b}{a}\right)$, if X is continuous,

2. $f_Y(y) = f_X\left(\dfrac{y-b}{a}\right)$, if X is discrete.

An important special case is that of the behavior of $aX + b$ when $X \sim N(\mu, \sigma^2)$.

Lemma 1.32 (Affine Transformations of Normal Distributions) *Let $X \sim N(\mu, \sigma^2)$, $a \neq 0$. Then $aX + b \sim N(a\mu + b, a^2\sigma^2)$. Consequently, if $X \sim N(\mu, \sigma^2)$, then*

$$F_X(x) = \Phi\left(\frac{x - \mu}{\sigma}\right),$$

where Φ is the standard normal CDF, $\Phi(u) = \int_{-\infty}^{u} (2\pi)^{-1/2} \exp\{-z^2/2\} dz$, that is, the distribution function of a random variable $Z \sim N(0,1)$.

Exercise 12 Prove Lemma 1.32.

This last result is particularly important because it allows us to calculate probabilities associated with normal random variables. The problem is that the integral $\int_{-\infty}^{u} \frac{1}{\sigma\sqrt{2\pi}} \exp\{-(x-\mu)^2/2\sigma^2\}dx$ cannot be explicitly solved, and so one would need to tabulate probabilities for all combinations of μ and σ (an impossible task). The last result tells us, however, that we only need to tabulate the standard normal CDF, Φ, and calculate probabilities by linear transformation. The process of subtracting the mean and then dividing by the standard deviation is called *standardisation*.

As a final result in this section, we state a theorem giving a general formula for the joint density of a bijective transformation of a collection of multiple random variables.

Theorem 1.33 (Multidimensional Transformations) *Let* $g : \mathbb{R}^n \to \mathbb{R}^n$ *be a continuously differentiable injection,*

$$g(x) = (g_1(x), \ldots, g_n(x)), \qquad x = (x_1, \ldots, x_n)^\top \in \mathbb{R}^n.$$

Let $X = (X_1, \ldots, X_n)^\top$ *be a random vector with joint density* $f_X(x)$, $x \in \mathbb{R}^n$, *and define* $Y = (Y_1, \ldots, Y_n)^\top = g(X)$. *Then, if* $\mathcal{Y}^n = g(\mathcal{X}^n)$, *we have*

$$f_Y(y) = f_X(g^{-1}(y)) \left| \det \left[J_{g^{-1}}(y) \right] \right|, \qquad \text{for } y = (y_1, \ldots, y_n)^\top \in \mathcal{Y}^n,$$

and zero otherwise, provided that $J_{g^{-1}}(y)$ *is well defined. Here,* $J_{g^{-1}}(y)$ *is the Jacobian of* g^{-1}, *i.e. the* $n \times n$-*matrix-valued function,*

$$J_{g^{-1}}(y) = \begin{bmatrix} \frac{\partial}{\partial y_1} g_1^{-1}(y) & \cdots & \frac{\partial}{\partial y_n} g_1^{-1}(y) \\ \vdots & \ddots & \vdots \\ \frac{\partial}{\partial y_1} g_n^{-1}(y) & \cdots & \frac{\partial}{\partial y_n} g_n^{-1}(y) \end{bmatrix}.$$

Exercise 13 Use the integration by substitution formula to prove the theorem.

Proposition 1.33 can sometimes be used in a clever way, even if the transformation involved is not invertible: it suffices to "augment" the transformation, as in the corollary that follows:

> **Corollary 1.34 (Convolution)** *Let X and Y be independent continuous random variables with densities f_X and f_Y. Then, the density of $X + Y$ is given by the convolution of f_X with f_Y:*
>
> $$f_{X+Y}(u) = \int_{-\infty}^{+\infty} f_X(u - v) f_Y(v) dv.$$

Proof To see this, define

$$g : \mathbb{R}^2 \to \mathbb{R}^2, \qquad (x, y) \overset{g}{\mapsto} (x + y, y)$$

with inverse mapping

$$(u, v) \overset{g^{-1}}{\mapsto} (u - v, v).$$

The Jacobian of the inverse can be easily seen to be

$$\begin{pmatrix} 1 & 0 \\ -1 & 1 \end{pmatrix}$$

whose absolute determinant is equal to 1. It follows from the multivariate transformation formula that

$$f_{X+Y,Y}(u, v) = f_{X,Y}(u - v, v) = f_X(u - v) f_Y(v),$$

where we have used the independence of X and Y. Integrating with respect to v now yields the marginal density f_{X+Y},

$$f_{X+Y}(u) - \int_{-\infty}^{+\infty} f_X(u - v) f_Y(v) dv.$$

$\square$

We conclude this section with an immediate application of the last corollary, concerning sums of normal random variables.

Corollary 1.35 (Sums of Independent Normal Random Variables) *Let $X_1, \ldots, X_n$ be independent random variables such that $X_i \sim N(\mu_i, \sigma_i^2)$, and let $S_n = \sum_{i=1}^n X_i$. Then,*

$$S_n \sim N\left(\sum_{i=1}^n \mu_i, \sum_{i=1}^n \sigma_i^2\right).$$

Proof It is clear that $\mathbb{E}[S_n] = \sum_{i=1}^n \mu_i$, so that we may assume that $\mu_i = 0$, and show that in this case $S_n \sim N(0, \sigma_1^2 + \cdots + \sigma_n^2)$. We proceed by induction, starting with $n = 2$. For tidiness, write $\sigma^2 = \sigma_1^2 + \sigma_2^2$. Then, by Corollary 1.34, we have

$$f_{X_1+X_2}(u) = \int_{-\infty}^{+\infty} f_X(u - v) f_Y(v) dv$$

$$= \int_{-\infty}^{+\infty} \frac{1}{\sigma_1\sigma_2 2\pi} \exp\left\{-\frac{\sigma_2^2 u^2 + \sigma_2^2 v^2 - 2\sigma_2^2 uv + \sigma_1^2 v^2}{2\sigma_1^2\sigma_2^2}\right\} dv.$$

Completing the square, we have

$$\sigma_2^2 u^2 + \sigma_2^2 v^2 - 2\sigma_2^2 uv + \sigma_1^2 v^2 = \sigma_2^2 u^2 + \sigma_2^2 v^2 - 2\sigma_2^2 uv$$

$$+ \sigma_1^2 v^2 + \sigma_2^4 \sigma^{-2} u^2 - \sigma_2^4 \sigma^{-2} u^2$$

$$= \left(\sigma_2^2 - \sigma_2^4 \sigma^{-2}\right) u^2 + \left(\sigma v - \sigma_2^2 \sigma^{-1} u\right)^2$$

$$\implies -\frac{\sigma_2^2 u^2 + \sigma_2^2 v^2 - 2\sigma_2^2 uv + \sigma_1^2 v^2}{2\sigma_1^2\sigma_2^2} = -\frac{u^2}{2\sigma^2} - \frac{\left(\sigma v - \sigma_2^2\sigma^{-1}u\right)^2}{2\sigma_1^2\sigma_2^2}.$$

Hence, with the change of variables $w = \sigma v$, we have

$$f_{X_1+X_2}(u) = \frac{1}{\sigma\sqrt{2\pi}} \exp\left\{-\frac{u^2}{2\sigma^2}\right\} \int_{-\infty}^{+\infty} \frac{\sigma}{\sigma_1\sigma_2\sqrt{2\pi}} \exp\left\{-\frac{\left(\sigma v - \sigma_2^2\sigma^{-1}u\right)^2}{2\sigma_1^2\sigma_2^2}\right\} dv$$

$$= \frac{1}{\sigma\sqrt{2\pi}} \exp\left\{-\frac{u^2}{2\sigma^2}\right\} \underbrace{\int_{-\infty}^{+\infty} \frac{1}{\sigma_1\sigma_2\sqrt{2\pi}} \exp\left\{-\frac{\left(w - \sigma_2^2\sigma^{-1}u\right)^2}{2\sigma_1^2\sigma_2^2}\right\} dw}_{=1}$$

since the integrand is the density of a Gaussian distribution with mean $\sigma_2^2\sigma^{-1}u$ and variance $\sigma_1^2\sigma_2^2$. In summary, we have

$$f_{X_1+X_2}(u) = \frac{1}{\sigma\sqrt{2\pi}} \exp\left\{-\frac{u^2}{2\sigma^2}\right\}$$

which is the density of a $N(0, \sigma^2)$ distribution.

For the induction step, suppose that we have proven that $S_k \sim N(0, \sigma_1^2 + \cdots + \sigma_k^2)$, and wish to prove that $S_{k+1} \sim N(0, \sigma_1^2 + \cdots + \sigma_{k+1}^2)$. Since

$$S_{k+1} = S_k + X_{k+1}$$

is the sum of a $N(0, \sigma_1^2 + \cdots + \sigma_k^2)$ with an independent $N(0, \sigma_{k+1}^2)$ random variable, the first part of the proof shows that indeed $S_{k+1} \sim N(0, \sigma_1^2 + \cdots + \sigma_{k+1}^2)$, and the proof is complete.   □

1.5   Model Selection and Exploratory Data Analysis

In the sequel, we will typically assume that a specific type of probability model has already been selected as a description of a random phenomenon, and will proceed in developing our theory taking this model as given. But before we do so, we must at least pause for a short moment and consider how or why such a model was selected in the first place. In other words, why does it make sense to assume that the exponential distribution is a good model for the waiting time until the emission of a radioactive particle, or the Poisson distribution in order to model the number of bacteria in a water tank? In very broad terms, we can say that the selection of a probability model could be based upon: (1) scientific theory and prior experimentation; (2) philosophical principles; (3) exploratory data analysis; (4) a combination of (1), (2) and (3).

The ideal situation is one where the modeler may choose a probability model as a consequence of a well-founded scientific theory or overwhelming empirical evidence. This is often the case in random phenomena that occur in the physical sciences, most commonly in physics, as a result of physical laws and/or experiments. These laws may suggest that the random phenomenon must satisfy certain conditions and/or possess some properties. If we are fortunate enough, we may have enough properties and conditions to uniquely determine a suitable probability model. There is much knowledge on whether or not a certain list of properties uniquely specifies a certain probability model in the field of *characterisation of probability models*.

Example 1.36 (Exponential Distribution for Emission Time)

Scientific theory suggests that it is impossible to predict how long it will take until an unstable nucleus decays. This time is a random variable T. In fact, the random process is such that even if a certain amount of time has elapsed and we have not yet seen a decay, this does not give us any information at all on how much longer we might still have to wait. In mathematical terms:

$$\mathbb{P}[T > t + s \,|\, T > t] = \mathbb{P}[T > s].$$

We know that the exponential distribution $f(t) = \lambda e^{-\lambda t} 1\{t > 0\}$ has this property. In fact, it can be proven that this is the only distribution supported on $[0, \infty)$ that has this property, thus dictating its choice in order to model radioactive particle emission times.   □

Exercise 14 Prove that the lack of memory property characterises the exponential. More precisely, let X be a random variable such that $\mathbb{P}(X > 0) > 0$ and

$$\mathbb{P}(X > t + s | X > t) = \mathbb{P}(X > s), \qquad \forall t, s \geq 0.$$

Prove that there exists a $\lambda > 0$ such that $X \sim \text{Exp}(\lambda)$.

  Hint: Let $G(t) = \mathbb{P}(X > t)$. Show that the lack of memory property implies that $G(t + s) = G(t)G(s)$ for $t, s \geq 0$. Then, define $g(t) = -\ln G(t)$ and $\lambda = g(1)$. Show that $g(t) = t\lambda$ for all $t > 0$ rational. Deduce that $g(t) = t\lambda$ for all $t \geq 0$. What is the sign of λ? Finally, show that $\lambda < \infty$ using the fact that $G(0) > 0$ and continuity from the right of G.

  It may seem that perfect characterisation of probability models is likely to occur only in relatively simple phenomena. This is not necessarily the case. Very often, we can build more and more complex models by combining several different constraints (stemming from theory or experiment), partial characterisations, approximations and mathematical manipulation. We will not consider here more elaborate examples, but will mention that Einstein's model for the movement of a particle in a gas or a liquid (the famous Brownian motion) can be developed by such means.

  Other times, even if we impose all the necessary conditions, we cannot uniquely determine a probability model. In other words, there are several candidate probability models that would respect the conditions imposed by scientific theory and experiment. If we have no other source of information or no other evidence to help us choose a model, then we might have to choose one by means of some sort of principle or postulate, for example a philosophical/epistemological principle.

Example 1.37 (Entropy)

Suppose that we wish to model a natural phenomenon whose outcome is described by a continuous random variable X taking values on a given $\mathcal{X} \subseteq \mathbb{R}$. Assume that scientific theory dictates that the phenomenon should satisfy certain properties on average, in the sense that the expectations of certain functions of X should be fixed:

$$\mathbb{E}[T_i(X)] = \alpha_i, \qquad i = 1, \ldots, k.$$

If there are several probability densities f under which X that would satisfy these expectation constraints, the philosophical principle of entropy dictates that among these we should prefer the model that maximises the entropy of X,

$$H(f) = -\int_{\mathcal{X}} \log[f(x)] f(x) dx.$$

The entropy of f is a measure of how "unpredictable" a random variable that follows f is. If we choose a density f that has low entropy, then we are in essence imposing a more"predictable" behaviour on X, a behaviour that is more favourable to us in terms of how easy it is to predict X. If we know nothing beyond our constraints, however, we do not wish to artificially impose such a simplification. We must therefore choose the worst case scenario, i.e. the most unpredictable model possible: the one that maximises the entropy.

A very interesting result says the following: if a maximiser of the entropy subject to the k expectation constraints exists, then it must be a k-parameter exponential family (in fact, the T_i that appear in the expectation constraints will also appear in the formula for the density of the specific exponential family). This explains why the exponential family features so prominently in probability models, and why the members of the exponential family form the fundamental examples used in much of statistics.   □

Example 1.38 (Parsimony)

If we are given two different probability models $f(\cdot; \theta)$ and $g(\cdot; \psi)$, depending on multi-dimensional parameters θ and ψ, respectively, both of which would satisfy equally well all the constraints and conditions that the random phenomenon should satisfy, choose the one that depends on the least effective number of parameters. For example, if θ can range in some d-dimensional set and ψ can range in some d'-dimensional set, with $d' < d$, we choose g over f. The principle of parsimony rests upon the idea that given different models that are adequate for the same phenomenon, we should choose the one that is least complex.   □

Still, there may be situations where a probability model cannot be unequivocally selected by means of physical laws and/or scientific principles, or where we are simply not willing to make a choice solely on the basis of a principle. In this case, we may seek out empirical evidence in order to supplement our principled choice of model, or in order to validate a model. For example, we may have observed n independent realisations of the random variable X. By looking at the characteristics of these n values we might be able to suggest a model that would appear fitting to the form of the data, or at minimum be able to rule out some models whose characteristics would be incompatible with what has been observed. The process of investigating patterns in the observed data in order to select an appropriate probability model is called *exploratory data analysis*.

1.5.1 Exploratory Data Analysis

Let $x_1, \ldots, x_n$ be a data set comprised of n real values. These values constitute the realisation of n independent and identically distributed random variables $X_1, \ldots, X_n$ whose probability distribution has a density/frequency function f which is unknown to us. Worse, still, we do not even know what class of distributions f belongs to. In order to be able to select an appropriate probability model, exploratory data analysis considers various graphical representations and numerical summaries of the data $x_1, \ldots, x_n$ that will allow us to gain an appreciation of the general form of f, along with some basic characteristics, that will hopefully guide our model choice.

What are some basic aspects of the form of a probability distribution that we can try to look for? Here are some of the most important characteristics that one ought to take into consideration:

1. **Location.** The location of a distribution is generally understood to be a point on the real line representing some centre of the distribution. The notion of a centre is a vague concept that can be made precise in several different ways. For example, it can be understood as a centre of mass (the mean, $\mu = \mathbb{E}[X]$), as a global maximum (the mode, $\arg\sup_{x \in \mathcal{X}} f(x)$), or a point that splits the probability mass in half (the median, $m = \inf\{x : F(x) \geq 1/2\}$). Notice that a location may not always be uniquely defined: though the mean is unique (when it exists), the mode may not be (e.g. think of a distribution with two peaks of equal height).

2. **Dispersion.** The dispersion of a distribution is a measure of how concentrated or diffuse the distribution is. Similarly to location, it can be formalised by several different measures. Often one measures dispersion by quantifying how concentrated the distribution is around a measure of location. For example, the variance $\mathbb{E}[(X - \mu)^2]$ is a classical measure of dispersion, that measures the second moment of inertia of the distribution around the mean. It is not the only one, though; for example, one may consider the *mean absolute deviation* (MAD), $\mathbb{E}[|X - \mu|]$, where $\mu = \mathbb{E}[X]$. Or further yet, one may consider a measure of dispersion that does not make explicit reference to the centre of a distribution. For example, the *interquartile range* is defined as $\text{IQR} = \inf\{x : F(x) \geq 3/4\} - \inf\{x : F(x) \geq 1/4\}$; roughly speaking, it measures the length of the most central interval supporting 50 % of the mass of the distribution.

3. **Symmetry/Skewness.** A density/frequency f is symmetric about a point x_0 if $f(x_0 - x) = f(x_0 + x)$ for all $x \in \mathcal{X}$. A distribution may be symmetric, mildly asymmetric (mildly skew) or strongly asymmetric (strongly skew). One may measure the asymmetry of a distribution through the notion of *skewness*, which is defined as: $\mathbb{E}\left[\left(\frac{X-\mu}{\sigma}\right)^3\right]$, where $\mu = \mathbb{E}[X]$ and $\sigma = \sqrt{\text{Var}[X]}$. If a distribution is symmetric, then its skewness must be zero. When the skewness is positive, we speak of a *right-skew distribution* (respectively, negative skewness yields a *left-skew distribution*).

4. **Tail Behaviour.** The tails of a distribution are the values taken by its density/frequency $f(x)$ as $x \to \pm\infty$. Notice that since f is always positive and integrates/sums to 1, it must be that $\lim_{x \to \infty} \mathbb{P}[|X| \geq x] = 0$. The rate of decay of $\mathbb{P}[|X| \geq x]$ as $x \to \infty$ is what determines its so-called tail behaviour. A light-tailed distribution has a fast rate of decay (for example, exponential), and a heavy tailed distribution has a slow rate of decay (for example, polynomial). A heavy tailed distribution is such that the probability of observing an extreme value is non-negligible. It might be that both the left and the right tails of a distribution are heavy, but it might also be that only one of these two tails is heavy (Fig. 1.9).

If a candidate probability model for a random variable X does not share similar location/dispersion/symmetry/tail properties as those observed for X, then it is not

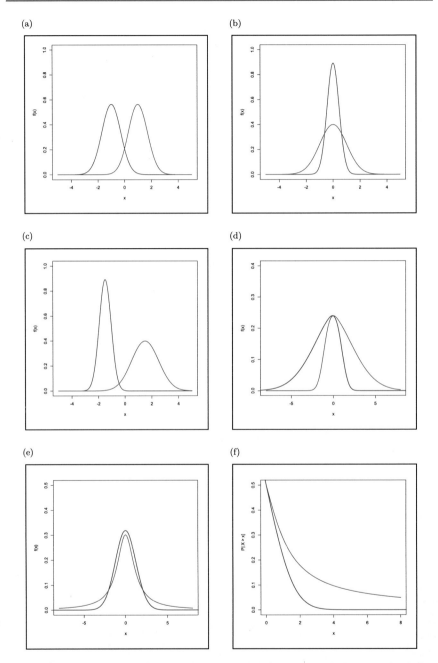

Fig. 1.9 Illustration of the notions of location, dispersion, skewness, and light/heavy tails. (**a**) Two densities differing in location. (**b**) Two densities differing in dispersion. (**c**) Two densities differing both in location and in dispersion. (**d**) Two asymmetric densities: one with positive skewness (*red*), and one with negative skewness (*blue*). (**e**) A heavy tailed density (*red*) and a light tailed density (*blue*). (**f**) Plots of the mapping $x \mapsto \int_x^\infty f(y)dy$ for the two densities on the left (**e**)

a good model for the phenomenon described by X. What do we mean by "those observed for X"? We mean that we can use the sample values $x_1, \ldots, x_n$ in order to gain some appreciation of these properties. We will do so *quantitatively* (using numerical summaries) and *qualitatively* (using graphical summaries).

1.5.1.1 Numerical Summaries

We first introduce some useful notation: if $x_1, \ldots, x_n$ are n real values, we denote by $x_{(j)}$ the jth sample value, when these are ordered in increasing ordered (so $x_{(1)} = \min\{x_1, \ldots, x_n\}$ and $x_{(n)} = \max\{x_1, \ldots, x_n\}$). Notice that this means that

$$x_{(1)} \leq x_{(2)} \leq \ldots \leq x_{(n-1)} \leq x_{(n)}.$$

To illustrate the notation, say that $n = 4$ and we have $x_1 = 5, x_2 = 12, x_3 = 2$, and $x_4 = 12$. Then we write $x_{(1)} = 2, x_{(2)} = 5$, and $x_{(3)} = x_{(4)} = 12$. So, in this case, $x_{(1)} = x_3, x_{(2)} = x_1, x_{(3)} = x_{(4)} = x_2 = x_4$.

With this notation under our belt, we begin by defining two numerical summaries of the sample that can be used in order to gauge the location of the sample.

Definition 1.39 (Sample Mean and Median)

Let $x_1, \ldots, x_n$ be a collection of real numbers, called a sample. We define:
1. The sample mean as

$$\bar{x} = \frac{1}{n} \sum_{i=1}^{n} x_i.$$

2. The sample median as

$$M = \begin{cases} x_{\left(\frac{n+1}{2}\right)} & \text{if } n \text{ is odd}, \\\\ \dfrac{x_{\left(\frac{n}{2}\right)} + x_{\left(\frac{n}{2}+1\right)}}{2} & \text{otherwise}. \end{cases}$$

Both of these characteristics have merits and drawbacks as descriptors of location. The mean takes into account the magnitude of each observation when determining location, and can be seen as the barycentre of the sample values.[2] However, it can be strongly affected by the presence of a single very large (or very small) value, which might distort the representativeness of the mean as a good descriptor of location. On the other hand, the median does not take into account

---

[2]That is, if we took the line segment $x_{(n)} - x_{(1)}$ and placed equal weights at the points $x_1, \ldots, x_n$, then the point $\bar{x}$ is where the line segment would balance.

the precise value of the observations, but simply their ordering, and can be seen as the "middle" observation[3] (or the average of the two middle observations, when the sample size is even). In this sense, it is a cruder indicator of location. This can also be an advantage, though: the median will not be sensitive to the presence of very large (or very small) observations, since it only takes their ordering (and not their magnitude) into account.

Exercise 15   1. Calculate the mean $\bar{x}$ and the median M of the following data set:

$$9.2 \quad 11.5 \quad 9.7 \quad 11.0 \quad 8.5$$
$$9.8 \quad 10.0 \quad 12.1 \quad 10.5 \quad 10.1$$

2. Repeat your calculation when the observation 12.1 is replaced by 48.6.
3. Compare the values of $\bar{x}$ and M in part 1 and part 2. What do you observe?

Exercise 16   Show that
1. The function $f(\gamma) = \sum_{i=1}^{n}(x_i - \gamma)^2$ has a unique minimum at $\bar{x}$.
2. The function $g(\gamma) = \sum_{i=1}^{n}|x_i - \gamma|$ is minimised at M. Warning: g is not differentiable at γ whenever $\gamma = x_i$ for $i = 1, \ldots, n$.

Next, we consider several numerical summaries that can be used in order to ascertain how disperse the underlying distribution might be on the basis of the sample values $x_1, \ldots, x_n$.

Definition 1.40 (Sample Variance and MAD)

Let $x_1, \ldots, x_n$ be a collection of real numbers, called a sample. We define:
1. The sample variance as

$$\hat{\sigma}^2 = \frac{1}{n}\sum_{i=1}^{n}(x_i - \bar{x})^2$$

(the sample standard deviation is defined as $\hat{\sigma} = \sqrt{\hat{\sigma}^2}$).
2. The sample MAD as

$$\text{MAD} = \frac{1}{n}\sum_{i=1}^{n}|x_i - \bar{x}|.$$

Exercise 17   Show that we may also write $\hat{\sigma}^2 = \frac{1}{n}\sum_{i=1}^{n}x_i^2 - \bar{x}^2$. Comment on why this formula may be more useful.

---

[3]In the sense that half the observations must be greater than or equal to the median, and half the observations must be less than or equal to the median.

The sample variance expresses how concentrated or spread out the observations are relative to their sample mean. From a physics point of view, it represents the second moment of inertia around the mean.[4] As was the case with the sample mean, the sample variance can also be substantially inflated when there is a single extreme observation in the sample. This will create an impression of much higher dispersion, when in fact the sample may be fairly well concentrated, with the exception of a single rogue observation. The MAD, on the other hand, is somewhat less affected in such circumstances, since it is formed by summing absolute distances, rather than squared distances (the square would disproportionally inflate the contribution of an extreme observation to the sum). One can show that when there are no extreme observations, the variance is a better indicator of dispersion; in the presence of extreme observations, the MAD is preferred. How can we judge which observations are extreme? The pertinent notion is that of *outliers*, whose presence is in fact an indicator of heavy tails.

Definition 1.41 (Quartiles, IQR and Outliers)
Let $x_1, \ldots, x_n$ be a sample of n real values, and let

$$x_{(1)}, \ldots, M, \ldots, x_{(n)}$$

be the ordered sample, where M is the median. We define:
1. The first quartile, Q_1, as the median of the ordered sub-sample $x_{(1)}, x_{(2)}, \ldots, M$.
2. The second quartile, Q_2 as being the median M, $Q_2 = M$.
3. The third quartile, Q_3, as the median of the ordered sub-sample $M, \ldots, x_{(n-1)}$, $x_{(n)}$.
4. The inter quartile range (IQR) as $\text{IQR} = Q_3 - Q_1$.
5. An outlier as an observation falling outside the interval $\left[Q_1 - \frac{3}{2}\text{IQR}, Q_3 + \frac{3}{2}\text{IQR}\right]$.

Just as the median can be interpreted as the "middle" observation, the first quartile can be seen as the "first quarter" observation (and the third quartile can be seen as the "third quarter" observation[5]). Half of the sample observations lie within the interval $[Q_1, Q_3]$. In some sense, the interval $[Q_1, Q_3]$ is the most central interval containing 50 % of the observations. The length of this interval, the IQR, can also be used as an indicator of dispersion. This length reflects how spread out the central portion of the sample is. Finally, the notions of quartiles and IQR can be used in order to define what would qualify as an "extreme" observation (an

---

[4]That is, if we took the line segment $x_{(n)} - x_{(1)}$ and placed equal weights at the points $x_1, \ldots, x_n$, then tried to rotate the segment around the point $\bar{x}$, then the variance is an indicator of how much force we would need to apply. If the observations are spread far from $\bar{x}$, then we need a lot of force (high sample variance); but if the observations are close to $\bar{x}$, then our task is easier (low sample variance).

[5]To be precise: 25 % of the sample observations are less than or equal to Q_1, and 25 % of the observations are greater than or equal to Q_3.

outlier). In some sense, extreme observations are removed from the bulk of the other observations. The definition of an outlier may seem somewhat arbitrary, but there are deeper mathematical reasons that support this definition.

Exercise 18 Let $x_1, \ldots, x_n$ be a sample. What are the median M and quartiles Q_1 and Q_3 when $n = 12, 13, 14$ or 15? A more tedious generalisation: find the general formulae (for n arbitrary) for the first and third quartile, Q_1 and Q_3. *Hint*: these formulae are of the form

$$\begin{cases} ? & n \equiv 0 \mod 4 \\ ? & n \equiv 1 \mod 4 \\ ? & n \equiv 2 \mod 4 \\ ? & n \equiv 3 \mod 4. \end{cases}$$

We conclude our brief discussion of numerical summaries by considering a measure of asymmetry: the *sample skewness*.

Definition 1.42 (Sample Skewness)

Let $x_1, \ldots, x_n$ be a sample of n real values. We define the skewness of this sample as

$$SK = \frac{\frac{1}{n} \sum_{i=1}^{n} (x_i - \bar{x})^3}{\left(\frac{1}{n} \sum_{i=1}^{n} (x_i - \bar{x})^2 \right)^{3/2}}.$$

If both the numerator and denominator are equal to zero (which can occur in discrete samples), then SK is undefined.

As was earlier discussed, one can look at whether SK is positive, negative, or close to zero, in order to judge whether the distribution generating the sample had a right or left asymmetry, or was indeed symmetric. A drawback is that the sample skewness may not be a good proxy for the true skewness of the distribution, and defining good bounds on "how large" the skewness should be in order to declare that the distribution is asymmetric is a subtle problem that requires methods from later chapters. Instead of embarking on such a project at this point, we turn to the use of graphical summaries, which will allow us to obtain an intuitive appreciation of the asymmetry in the data, without needing to resort to elaborate calculations.

1.5.1.2 Graphical Summaries

We now turn to two simple graphical representations of the sample $x_1, \ldots, x_n$ that can help us visualise the form of the underlying density/frequency f. The *histogram* and the *boxplot*. A histogram is a proxy for the unknown density built out of the observed sample values $x_1, \ldots, x_n$. The idea is simple: if there are many observations falling in some interval I, then the density should be relatively high on that interval. Therefore, if we partition the x-axis into disjoint intervals, and define a step function that is constant over these intervals (and such that the height of each step is proportional to the percentage of observations lying in the corresponding interval), we will have constructed a step function approximation to the unknown density.

Definition 1.43 (Histogram)

Let $x_1, \ldots, x_n$ be a collection of n real values and $h > 0$ be a constant. Let $\{I_j\}_{j \in \mathbb{Z}}$ be a regular partition of $\mathbb{R}$ comprised of intervals of length $h > 0$,

$$I_j = \Big[\kappa + (j - 1)h, \kappa + jh\Big), \qquad j \in \mathbb{Z},$$

where $\kappa \in \mathbb{R}$ is some fixed real number. The histogram of $x_1, \ldots, x_n$ with bin width $h > 0$ and origin κ is defined to be the graph of the function:

$$y \;\mapsto\; \mathrm{hist}_{x_1, \ldots, x_n}(y) = \frac{1}{h} \sum_{j \in \mathbb{Z}} \mathbf{1}\{y \in I_j\} \frac{1}{n} \sum_{i=1}^{n} \mathbf{1}\{x_i \in I_j\}.$$

Notice that the histogram is indeed a reasonable step function approximation of f: by its definition, the function $\mathrm{hist}_{x_1, \ldots, x_n}(y)$ takes non-negative values only, and the integral of the function $\mathrm{hist}_{x_1, \ldots, x_n}(y)$ is equal to 1. In addition, the integral of $\mathrm{hist}_{x_1, \ldots, x_n}(y)$ over an interval I_j gives us the proportion of sample values that fell inside I_j. It therefore has the properties of a probability density function. Furthermore,

$$\mathbb{E}\left[\int_{I_j} \mathrm{hist}_{X_1, \ldots, X_n}(y)dy\right] = \frac{1}{n} \sum_{i=1}^{n} \mathbb{P}[X_i \in I_j] = \int_{I_j} f(y)dy.$$

In this sense, the histogram is some sort of Riemann-sum-proxy of the density f, constructed using the values of the sample. It can be used in order to gauge properties such as location, dispersion, symmetry and tail behaviour via a visual inspection.

▶ **Remark 1.44 (Bin Width)**   Depending on the choice of h a histogram may be more or less informative about the structure of the sample at hand. Consider the two extremes, $h \to 0$ and $h \to \infty$. In the first case, the intervals eventually become so short that any interval contains either no observations or a single observation, thus simply highlighting where each observation lies on the x-axis (see Fig. 1.10e, p. 36). In the second case, all the observations are eventually contained in a single huge interval, and the histogram simply informs us that there is a large region that contains all observations (see Fig. 1.10f, p. 36). Reasonable values of h allow us to visualise the structure of the sample. In principle, the value of h should depend on

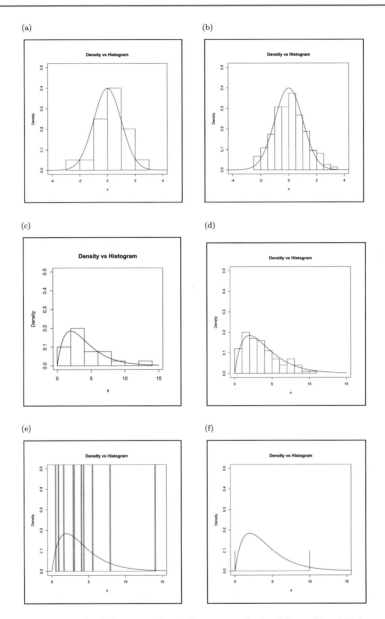

Fig. 1.10 Histograms for different samples (and, correspondingly, different bin widths) compared with the density from which the samples were drawn. (**a**) Density of $N(0, 1)$ (in *red*) and histogram for a random sample of size 20 from an $N(0, 1)$ (in *black*). (**b**) Density of $N(0, 1)$ (in *red*) and histogram for a random sample of size 100 from an $N(0, 1)$ (in *black*). (**c**) Density of χ_2^2 (in *red*) and histogram for a random sample of size 20 from a χ_2^2 (in *black*). (**d**) Density of χ_2^2 (in *red*) and histogram for a random sample of size 100 from a χ_2^2 (in *black*). (**e**) Density of χ_2^2 (in *red*) and histogram for a random sample of size 20 from a χ_2^2 (in *black*) when the bin width h is taken to be very small. (**f**) Density of χ_2^2 (in *red*) and histogram for a random sample of size 20 from a χ_2^2 (in *black*) when the bin width h is taken to be very large

the sample size n: the larger n, the smaller h needs to be; intuitively, this means that when we have more observations, we can try to investigate finer aspects of the structure of the sample $x_1, \ldots, x_n$. The precise requirement is that we must have that $h \overset{n \to \infty}{\longrightarrow} 0$ and $hn \overset{n \to \infty}{\longrightarrow} \infty$. There is a lot of theory on what the optimal h is as dependent on n, but we will not consider this here. A simple (but often suboptimal) choice is to take $h = n^{-1/2}$. A data-dependent choice is the so-called *Freedman–Diaconis* choice of $h = 2\text{IQR} \times n^{-\frac{1}{3}}$.

▶ **Remark 1.45 (Bin Centres)**   Notice that for any given $h > 0$, there are several possible histograms depending on the choice of κ. Unfortunately, there is no unequivocal means of determining what the "right" κ is. The analyst must either try several values or at minimum keep in mind that the histogram should not be over-interpreted, as its form may be perturbed by changes in κ (e.g. because by a small shift in κ some observations that fell in the kth interval may now fall in the $(k + 1)$th interval, and so on).

Histograms can be criticised for having some noticeable drawbacks. Chief among these is the need to choose a bin width h and an origin κ. Another drawback is that they can sometimes become misleading if over-interpreted. For example, looking at the histogram in Fig. 1.10c (p. 36) we see a slight pattern of asymmetry. Is this to be taken as an indication that the underlying distribution is asymmetric? Not necessarily, since a histogram can rarely be perfectly symmetric due to sampling variation. The message here is that we should not try to extract finer information than what our graphical summary is really able to offer. Histograms may deceivingly appear to be interpretable in more detail than they actually are.

A different type of graphical display that allows us to probe the location, scale, asymmetry and tails of a density is the boxplot. In contrast to the histogram, the boxplot is a much coarser description of the sample structure and does not require the specification of any tuning parameters. It simply marks out the points on the x-axis where some key numerical summaries of a sample are located. This is usually done in the form of a box, which explains the name of the boxplot:

Definition 1.46 (Boxplot)
Let $x_1, \ldots, x_n$ be a collection of n real values. Let:
1. M be the median, Q_1 be the first quartile, and Q_3 be the third quartile of $\{x_1, \ldots, x_n\}$.
2. $W_1 = \min_{1 \le j \le n} \{x_j : x_j \ge Q_1 - 1.5 \times \text{IQR}\}$ & $W_2 = \max_{1 \le j \le n} \{x_j : x_j \le Q_3 + 1.5 \times \text{IQR}\}$.
3. $O = \{i \in \{1, \ldots, n\} : x_i \notin [W_1, W_2]\}$.
   The boxplot of $x_1, \ldots, x_n$ is an annotation of the values M, Q_1, Q_3, W_1, W_2, and $\{x_j : j \in O\}$ on the real line. The following is a standard annotation:

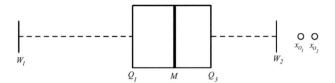

The definition is a little difficult to visualise, but the picture says it all: we annotate the median (M), the first and third quartiles (Q_1 and Q_3), and the first and last observation (W_1 and W_2) to fall within the interval $[Q_1 - 1.5 \times \text{IQR}, Q_3 + 1.5 \times \text{IQR}]$ (these two observations are called the *whiskers*). Any observations falling outside of the whiskers are marked separately and are *outliers* (the $\{x_j : j \in O\}$). Since $W_1 \leq Q_1 \leq M \leq Q_3 \leq W_2$, we usually omit the explicit annotation, since by their ordering it is clear which component of the boxplot denotes which value.

The boxplot illustrates the location of the sample by means of the median. It also gives an indication of the underlying dispersion by presenting the quartiles Q_1 and Q_3 (and their distance) as well as the whiskers (W_1 and W_2). Large distances between these values indicate large dispersion. Asymmetries can be probed by looking at the positioning of the quartiles and of the whiskers relative to the median. If these are located roughly symmetrically opposite of the median on either side, then we have a roughly symmetric structure. If the distance of one of the quartiles or one of the whiskers from the median is greater than that of the other, then we have skewness towards the side where the distance is greater. Finally a boxplot allows us to detect the presence of heavy tails, by looking at how many outliers there are, and on which tail of the distribution these are. Again, it is easiest to appreciate different forms of boxplots by looking at some pictures (see Fig. 1.11, p. 39).

Exercise 19 The following data are on the maximal weight (in tons) that could be supported by steel cables produced at a factory:

$$
\begin{array}{ccccccc}
10.1 & 12.2 & 9.3 & 12.4 & 13.7 & 11.1 & 13.3 \\
10.8 & 11.6 & 10.1 & 11.2 & 11.4 & 11.8 & 7.1 \\
12.2 & 12.6 & 9.2 & 14.2 & 10.5 & &
\end{array}
$$

1. Represent the data in a histogram with bin width $h = 1$ and origin $\kappa = 10$. Construct a second histogram, this time with $h = 2$ and $\kappa = 11$ and compare the two.
2. What is the approximate weight that at least 3/4 of the cables can support?
3. Find the third quartile.
4. Construct a box plot. Are there any outliers to be noticed? Where does one find the value determined in part (2) in this diagram?

Exercise 20 The following table contains the results of rugby matches of the eleventh and twelfth match days (November 2014) of the French rugby first ("Top 14") and second ("Pro D2") division. The home team is always mentioned first.

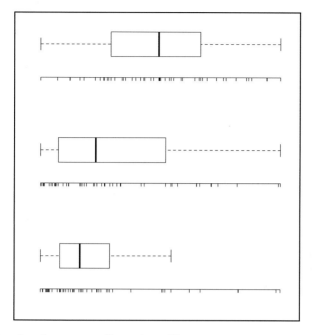

Fig. 1.11 Three boxplots corresponding to three different samples. In each case, the ticks on the axis below the boxplot represent the actual sample values from which the boxplot was constructed. Some noticeable aspects of the three samples based on the boxplots are: the first sample seems to present a high degree of symmetry. Both of the remaining samples show a clear asymmetry, and they are both skewed to *the right* (positive skewness). The third sample appears to present heavy right tails, as indicated by the presence of multiple outliers

Top 14		D2	
Montpellier–Brive	10–25	Albi–Agen	22–9
Castres–Toulon	22–14	Béziers–Aurillac	14–19
Clermont–Stade Français	51–9	Colomiers–Pau	50–10
Grenoble–Lyon	34–30	Montauban–Tarbes	31–13
Oyonnax–La Rochelle	37–9	Biarritz–Massy	21–3
Racing Métro–Bayonne	27–10	Dax–Narbonne	12–3
Bordeaux Bègles–Toulouse	20–21	Perpignan–Bourgoin	42–0
		Carcassonne–Mont-de-Marsan	17–28
Toulon–Clermont	27–19	Biarritz–Agen	42–18
Castres–Racing Métro	9–14	Albi–Carcassonne	34–22
La Rochelle–Bayonne	19–19	Aurillac–Colomiers	20–13
Lyon–Montpellier	23–20	Bourgoin–Montauban	14–20
Oyonnax–Bordeaux Bègles	28–23	Massy–Dax	50–13
Toulouse–Grenoble	22–25	Mont-de-Marsan–Béziers	32–18
Stade Français–Brive	20–17	Narbonne–Tarbes	36–23
		Pau–Perpignan	22–19

1. We wish to compare the performance of the teams in the first and second division. To this aim, calculate the pertinent statistics (mean, median, quartiles, EQR, etc.) for the score difference, as well as for the total points scored in each match, for each of the two division.
2. Construct box plots, and juxtapose them, for the sum and difference of points in each division, respectively. What conclusions can we draw?

Sampling from Probability Distributions

2

As mentioned in the introduction, statistical inference deals with the problem of making inferences from data in the presence of uncertainty. The mathematical framework for this endeavour is provided by probability models. At a general level, an inferential task can be cast as:

1. A random phenomenon X is assumed to be described by a regular parametric probability model $\{F_\theta : \theta \in \Theta\}$. The functional form of each F_θ is completely known, for any value of the parameter $\theta \in \Theta \subseteq \mathbb{R}^p$.
2. We observe a sample from a specific version of this probability model. That is, we observe n independent and identically distributed realisations $X_1, \ldots, X_n$ having distribution $F(x; \theta)$, for some $\theta \in \Theta$. Though we know that our observations stem from a version of the parametric regular model, we do not know the precise θ that generated the data (i.e. we know the model, but we do not know which member of the model generated the data).
3. We wish to use the sample $(X_1, \ldots, X_n)$ at hand in order to make statements about the true value of θ that generated it, and quantify the uncertainty attached to those statements.

2.1 Sampling, Statistics and Sufficiency

Since the sample is all we have, anything we do will essentially be a function of the sample, say $T(X_1, \ldots, X_n)$. Such a function is called a *statistic*.

Definition 2.1 (Statistic)
Let $\mathcal{X}$ be a sample space. Given $n \geq 1$, a statistic is a function $T : \mathcal{X}^n \to \mathbb{R}$.

Notice that the function T cannot depend on the parameter θ, since we do not know the latter. If the function T also depends on θ, it cannot be called a statistic.

Since a statistic $T : \mathcal{X}^n \to \mathbb{R}$ reduces a collection of n numbers to a single number, it cannot be injective. As a result $T(X_1, .., X_n)$ will in general provide less

© Springer International Publishing Switzerland 2016
V.M. Panaretos, *Statistics for Mathematicians*, Compact Textbooks in Mathematics,
DOI 10.1007/978-3-319-28341-8_2

information about θ than the complete data $(X_1, \ldots, X_n)$ will. For some models, however, we are able to choose a statistic T such that $T(X_1, \ldots, X_n)$ is equally informative about θ as $(X_1, \ldots, X_n)$ is. Such a statistic is called a *sufficient statistic* (because it suffices to use $T(X_1, \ldots, X_n)$ in lieu of $(X_1, \ldots, X_n)$).

Definition 2.2 (Sufficiency)

Let $X_1, \ldots, X_n \overset{iid}{\sim} f_\theta$. A statistic $T : \mathcal{X}^n \to \mathbb{R}$ is called sufficient for the parameter θ, if $\mathbb{P}[X_1 \leq x_1, \ldots, X_n \leq x_n | T = t]$ does not depend on θ, for all $(x_1, \ldots, x_n)^\top \in \mathbb{R}^n$ and all $t \in \mathbb{R}$.

The intuitive interpretation of this definition is: given the value of $T(X_1, \ldots, X_n)$, the conditional distribution of $(X_1, \ldots, X_n)$ no longer depends on θ. Therefore, knowing $(X_1, \ldots, X_n)$ in addition to knowing $T(X_1, \ldots, X_n)$ cannot furnish any more or any less information about which θ generated the data. The definition is usually hard to verify, but the following equivalent condition is much easier to verify:

Theorem 2.3 (Fisher–Neyman Factorisation) *Suppose that $(X_1, \ldots, X_n)$ has a joint density/frequency function $f_{X_1, \ldots, X_n}(x_1, \ldots, x_n; \theta)$, $\theta \in \Theta$. A statistic $T : \mathcal{X}^n \to \mathbb{R}$ is sufficient for θ if and only if there exist $g : \mathbb{R} \times \Theta \to \mathbb{R}$ and $h : \mathcal{X}^n \to \mathbb{R}$ such that*

$$f_{X_1, \ldots, X_n}(x_1, \ldots, x_n; \theta) = g(T(x_1, \ldots, x_n), \theta)h(x_1, \ldots, x_n).$$

Proof The proof in the continuous case requires the use of measure theory. Therefore, we will only give the proof in the case where the X_i are discrete random variables. Notice that if the X_i are discrete, then $T(X_1, \ldots, X_n)$ must also be discrete. Suppose that T is sufficient. Then,

$$
\begin{aligned}
f_{X_1, \ldots, X_n}(x_1, \ldots, x_n; \theta) &= \mathbb{P}_\theta[X_1 = x_1, \ldots, X_n = x_n] \\
&= \mathbb{P}_\theta[X_1 = x_1, \ldots, X_n = x_n, T = T(x_1, \ldots, x_n)] \\
&\quad + \underbrace{\mathbb{P}_\theta[X_1 = x_1, \ldots, X_n = x_n, T \neq T(x_1, \ldots, x_n)]}_{=0} \\
&= \mathbb{P}_\theta[T = T(x_1, \ldots, x_n)] \mathbb{P}_\theta[X_1 = x_1, \ldots, X_n = x_n | T = T(x_1, \ldots, x_n)]
\end{aligned}
$$

Since T is sufficient, the second term is independent of θ and so the Fisher–Neyman factorisation follows. To prove the converse, suppose that $f_{X_1, \ldots, X_n}(x_1, \ldots, x_n; \theta) =$

$g(T(x_1, \ldots, x_n), \theta)h(x_1, \ldots, x_n)$. Then,

$$\mathbb{P}_\theta[X_1 = x_1, \ldots, X_n = x_n | T = t]$$

$$= \frac{\mathbb{P}_\theta[X_1 = x_1, \ldots, X_n = x_n, T = t]}{\mathbb{P}_\theta[T = t]}$$

$$= \frac{\mathbb{P}_\theta[X_1 = x_1, \ldots, X_n = x_n]}{\mathbb{P}_\theta[T = t]} \mathbf{1}\{T(x_1, \ldots, x_n) = t\}$$

$$= \frac{\mathbb{P}_\theta[X_1 = x_1, \ldots, X_n = x_n]\mathbf{1}\{T(x_1, \ldots, x_n) = t\}}{\sum_{y_1 \in \mathscr{X}} \cdots \sum_{y_n \in \mathscr{X}} \mathbb{P}_\theta[X_1 = y_1, \ldots, X_n = y_n]\mathbf{1}\{T(y_1, \ldots, y_n) = t\}}$$

$$= \frac{g(T(x_1, \ldots, x_n); \theta)h(x_1, \ldots, x_n)\mathbf{1}\{T(x_1, \ldots, x_n) = t\}}{\sum_{y_1 \in \mathscr{X}} \cdots \sum_{y_n \in \mathscr{X}} g(T(y_1, \ldots, y_n); \theta)h(y_1, \ldots, y_n)\mathbf{1}\{T(y_1, \ldots, y_n) = t\}}$$

$$= \frac{g(t; \theta)h(x_1, \ldots, x_n)\mathbf{1}\{T(x_1, \ldots, x_n) = t\}}{g(t; \theta)\sum_{y_1 \in \mathscr{X}} \cdots \sum_{y_n \in \mathscr{X}} h(y_1, \ldots, y_n)\mathbf{1}\{T(y_1, \ldots, y_n) = t\}}$$

$$= \frac{h(x_1, \ldots, x_n)\mathbf{1}\{T(x_1, \ldots, x_n) = t\}}{\sum_{y_1 \in \mathscr{X}} \cdots \sum_{y_n \in \mathscr{X}} h(y_1, \ldots, y_n)\mathbf{1}\{T(y_1, \ldots, y_n) = t\}}.$$

and the latter does not depend on θ because neither h (by its definition) nor T (being a statistic) depend on θ. □

Example 2.4 (Estimating the Bias of a Coin)

Let $X_1, \ldots, X_n \overset{iid}{\sim} Bern(p)$. Then,

$$f_{X_1, \ldots, X_n}(x_1, \ldots, x_n) = \prod_{i=1}^n f_{X_i}(x_i) = p^{\left(\sum_{i=1}^n \mathbf{1}\{x_i=1\}\right)}(1-p)^{\left(n-\sum_{i=1}^n \mathbf{1}\{x_i=1\}\right)}.$$

Therefore, the Fisher–Neyman factorisation is satisfied with $T(X_1, \ldots, X_n) = \sum_{i=1}^n \mathbf{1}\{X_i = 1\} = \sum_{i=1}^n X_i$ (the last equality is because each X_i is 0 or 1), $g(t, p) = p^t(1-p)^{n-t}$ and $h(x_1, \ldots, x_n) = 1$. It follows that $\sum_{i=1}^n X_i$ is sufficient for p. Intuitively: knowing the total number of heads is all that matters as far as learning about p. Knowing the precise order in which these heads came up is irrelevant as far as p is concerned. □

When applied to a sample, any statistic (whether sufficient or not) becomes itself a random variable, one that has a distribution of its own. This is called a *sampling distribution*, because it arises as the result of random sampling.

Definition 2.5 (Sampling Distribution)

Let $X_1, \ldots, X_n \overset{iid}{\sim} F$ and $T : \mathcal{X}^n \to \mathbb{R}$ be a statistic. The sampling distribution of T under the distribution F is the probability distribution

$$F_T(t) = \mathbb{P}[T(X_1, \ldots, X_n) \leq t], \qquad t \in \mathbb{R}.$$

▶ **Remark 2.6 (Notation)** We always consider statistics as applied to a sample, and so we will very often suppress the dependence of the statistic on $X_1, \ldots, X_n$, and write simply T instead of $T(X_1, \ldots, X_n)$. In this notation, the sampling distribution of T under F is $F_T(t) = \mathbb{P}[T \leq t]$.

Exercise 21 Let $X_1, \ldots, X_n \overset{iid}{\sim} Unif(0, \theta)$. Show that $T(X_1, \ldots, X_n) = X_{(n)}$ is a sufficient statistic for θ, and find its sampling distribution.

Exercise 22 Let $X_1, \ldots, X_n \overset{iid}{\sim} Pois(\lambda)$. Show that $T(X_1, \ldots, X_n) = \sum_{i=1}^{n} X_i$ is a sufficient statistic for λ, and find its sampling distribution

Note that in the definition of the sampling distribution of T we specified under which distribution it occurs. This needs to be done, since changing the distribution of $X_1, \ldots, X_n$ to some G instead of F will also change the sampling distribution of T. In this chapter we will investigate precisely the dependence of this sampling distribution on the form of T and the form of F. Specifically:

- We will investigate some special forms of T and some special cases of F where the sampling distribution is known exactly.
- In more general situations, when the form of T and F might not allow for a straightforward determination of the sampling distribution, we will try to give ways of establishing an approximate distribution (and the mathematical framework required to make sense of "approximate distribution").

The statistics T that we will focus on will be sufficient statistics, and the models F will be members of exponential families.

2.2   Sampling from a Normal Distribution

We begin with the simplest possible problem: establishing the sampling distribution of the statistics

$$\bar{X} = \frac{1}{n} \sum_{i=1}^{n} X_i \quad \& \quad S^2 = \frac{1}{n-1} \sum_{i=1}^{n} (X_i - \bar{X})^2$$

when the sample $X_1, \ldots, X_n$ is a random sample from the normal distribution, i.e. $X_1, \ldots, X_n \overset{iid}{\sim} N(\mu, \sigma^2)$. Note that $\bar{X}$ is simply the empirical mean, while S^2 is $n/(n-1)$ times the empirical variance (the reason for using S^2 instead of the empirical variance will be seen very shortly). Though this problem seems relatively elementary, we will see that, for many other distributions, and for many other types of statistics, we can reduce the problem of determining the sampling distribution of those statistics to (approximately) a problem involving empirical means and variances of (approximately) normal random variables. We summarise the sampling distribution of $\bar{X}$ and S^2 in the next proposition.

Proposition 2.7 (Gaussian Sampling) *Let* $X_1, \ldots, X_n \overset{iid}{\sim} N(\mu, \sigma^2)$. *Then,*
1. *The joint distribution of* $X_1, \ldots, X_n$ *has probability density function,*

$$f_{X_1, \ldots, X_n}(x_1, \ldots, x_n) = \left(\frac{1}{2\pi\sigma^2}\right)^{n/2} \exp\left\{-\frac{1}{2\sigma^2} \sum_{i=1}^{n} (x_i - \mu)^2\right\}.$$

2. *The sample mean satisfies* $\bar{X} \sim N(\mu, \sigma^2/n)$.
3. *The random variables* $\bar{X}$ *and* S^2 *are independent.*
4. *The random variable* S^2 *satisfies* $\dfrac{n-1}{\sigma^2} S^2 \sim \chi^2_{n-1}$.

Proof For part (1), it suffices by independence to take the product of the marginal $N(\mu, \sigma^2)$ densities in order to arrive at the expression for the joint density.

For part (2), the fact that the random variables $X_1, \ldots, X_n$ are independent normal variables implies that $\sum_{i=1}^{n} X_i$ is also a normal random variable, with mean $n\mu$ and variance $n\sigma^2$ (by Corollary 1.35, p. 25). It follows that $\bar{X} = n^{-1} \sum_{i=1}^{n} X_i \sim N(\mu, \sigma^2/n)$.

For part (3), we note that if we can prove the independence of $\bar{X}$ from $X_1 - \bar{X}, \ldots, X_n - \bar{X}$ then it will immediately follow that $\bar{X}$ and S^2 are independent. To show this, write

$$Y_1 = \bar{X} \quad \& \quad Y_j = X_j - \bar{X}, \quad j = 2, \ldots, n.$$

Notice that the transformation $(X_1, \ldots, X_n) \mapsto (Y_1, \ldots, Y_n)$ is a linear bijection $\mathbb{R}^n \to \mathbb{R}^n$ because

$$\begin{aligned} Y_1 &= \bar{X} & X_1 &= Y_1 - \sum_{i=2}^{n} Y_i \\ Y_2 &= X_2 - \bar{X} & X_2 &= Y_2 + Y_1 \\ Y_3 &= X_3 - \bar{X} & X_3 &= Y_3 + Y_1 \\ &\vdots & &\vdots \\ Y_n &= X_n - \bar{X} & X_n &= Y_n + Y_1 \end{aligned}$$

Since the transformation is linear, its Jacobian is a constant that does not depend on $(X_1, .., X_n)$ (it is in fact equal to $1/n$). It follows from our results on transformations of random variables (Theorem 1.33, p. 23) that the joint density of $(Y_1, \ldots, Y_n)$ is given by

$$f_{Y_1,\ldots,Y_n}(y_1,\ldots,y_n) = n f_{X_1,\ldots,X_n}(x_1,\ldots,x_n)$$

$$= \frac{n}{(2\pi\sigma^2)^{n/2}} \exp\left\{-\frac{1}{2}\sum_{i=1}^{n}\left(\frac{x_i - \mu}{\sigma}\right)^2\right\}$$

$$= \frac{n}{(2\pi\sigma^2)^{n/2}} \exp\left\{-\frac{1}{2}\sum_{i=1}^{n}\left(\frac{x_i - \bar{x} + \bar{x} - \mu}{\sigma}\right)^2\right\}.$$

But $\sum_{i=1}^{n}(x_i - \bar{x}) = \sum_{i=1}^{n} x_i - n\bar{x} = n\bar{x} - n\bar{x} = 0$ so, on the one hand, $\sum_{i=1}^{n}(x_i - \bar{x})(\bar{x} - \mu) = 0$ and, on the other hand, $(x_1 - \bar{x}) = -\sum_{i=2}^{n}(x_i - \bar{x})$. Using these two identities gives us:

$$f_{Y_1,\ldots,Y_n}(y_1,\ldots,y_n)$$

$$= \frac{n}{(2\pi\sigma^2)^{n/2}} \exp\left\{-\frac{1}{2\sigma^2}\left(\sum_{i=1}^{n}(x_i - \bar{x})^2 + n(\bar{x} - \mu)^2\right)\right\}$$

$$= \frac{n}{(2\pi\sigma^2)^{n/2}} \exp\left\{-\frac{1}{2\sigma^2}\left((x_1 - \bar{x})^2 + \sum_{i=2}^{n}(x_i - \bar{x})^2 + n(\bar{x} - \mu)^2\right)\right\}$$

$$= \frac{n}{(2\pi\sigma^2)^{n/2}} \exp\left\{-\frac{1}{2\sigma^2}\left[\left(\sum_{i=2}^{n}(x_i - \bar{x})\right)^2 + \sum_{i=2}^{n}(x_i - \bar{x})^2 + n(\bar{x} - \mu)^2\right]\right\}$$

$$= \frac{n}{(2\pi\sigma^2)^{n/2}} \exp\left\{-\frac{1}{2\sigma^2}\left[\left(\sum_{i=2}^{n}y_i\right)^2 + \sum_{i=2}^{n}y_i^2 + n(y_1 - \mu)^2\right]\right\}$$

$$= \underbrace{\frac{\sqrt{n}}{(2\pi\sigma^2)^{(n-1)/2}} \exp\left\{-\frac{1}{2\sigma^2}\left[\left(\sum_{i=2}^{n}y_i\right)^2 + \sum_{i=2}^{n}y_i^2\right]\right\}}_{f_1(y_2,\ldots,y_n)}$$

$$\underbrace{\frac{1}{(2\pi\sigma^2/n)^{1/2}} \exp\left\{-\frac{1}{2\sigma^2/n}\left[(y_1 - \mu)^2\right]\right\}}_{f_2(y_1)}$$

Notice that $f_2(y_1)$ is the marginal density of $Y_1 = \bar{X} \sim N(\mu, \sigma^2/n)$, as proven in part (2). Therefore, if we integrate both sides with respect to y_1, we obtain that $f_2(y_2,\ldots,y_n)$ is the joint density of $(Y_2,\ldots,Y_n)$. We thus conclude that

$$f_{Y_1,\ldots,Y_n}(y_1,\ldots,y_n) = f_{Y_1}(y_1) f_{Y_2,\ldots,Y_n}(y_2,\ldots,y_n).$$

Consequently, $Y_1 = \bar{X}$ is independent of $Y_2 = X_2 - \bar{X},\ldots, Y_n = X_n - \bar{X}$. Since $(X_1 - \bar{X}) = -\sum_{i=2}^{n}(X_i - \bar{X})$, it follows that Y_1 is also independent of $X_1 - \bar{X}$. This proves (3), i.e. that $\bar{X}$ and S^2 are independent.

To prove (4) we note that

$$\sum_{i=1}^{n}(X_i - \mu)^2 = \sum_{i=1}^{n}(X_i - \bar{X})^2 + 2\underbrace{\sum_{i=1}^{n}(X_i - \bar{X})(\bar{X} - \mu)}_{=0} + n(\bar{X} - \mu)^2$$

$$= (n-1)S^2 + n(\bar{X} - \mu)^2$$

$$\Longrightarrow \underbrace{\sum_{i=1}^{n}\left(\frac{X_i - \mu}{\sigma}\right)^2}_{Q} = \underbrace{\frac{(n-1)}{\sigma^2}S^2}_{V} + \underbrace{\left(\frac{\bar{X} - \mu}{\sigma/\sqrt{n}}\right)^2}_{W}$$

Since we have proven in part (3) that S^2 and $\bar{X}$ are independent, it follows that the MGF of Q must be the product of the MGFs of V and of W (again by Lemma A.10, p. 168):

$$M_Q(t) = M_V(t)M_W(t).$$

From part (2) we know that $\frac{\bar{X}-\mu}{\sigma/\sqrt{n}} \sim N(0, 1)$ and thus $W \sim \chi_1^2$ (as the square of a standard normal random variable, see Eq. (1.4), p. 21) and so

$$M_W(t) = (1 - 2t)^{-1/2}.$$

We also know that $\frac{X_i-\mu}{\sigma} \overset{iid}{\sim} N(0, 1)$, so it is also true that $\left(\frac{X_i-\mu}{\sigma}\right)^2 \overset{iid}{\sim} \chi_1^2$. Therefore, the MGF of Q is equal to:

$$M_Q(t) = \prod_{i=1}^{n}(1 - 2t)^{-1/2} = (1 - 2t)^{-n/2}.$$

Summarising, we have that

$$\underbrace{(1 - 2t)^{-n/2}}_{M_Q(t)} = M_V(t)\underbrace{(1 - 2t)^{-1/2}}_{M_W(t)},$$

from which it follows that

$$M_V(t) = (1 - 2t)^{-(n-1)/2}.$$

This is the MGF of the χ_{n-1}^2 distribution. Since the MGF completely determines a distribution (Proposition A.9, p. 165), this proves part (4) and completes the proof. □

The following follows immediately from the theorem:

Corollary 2.8 (Moments for Normal Sampling) *Let* $X_1, \ldots, X_n \overset{iid}{\sim} N(\mu, \sigma^2)$. *Then,*

$$\mathbb{E}[\bar{X}] = \mu, \quad \mathrm{Var}(\bar{X}) = \frac{\sigma^2}{n}, \quad \mathbb{E}[S^2] = \sigma^2, \quad \mathrm{Var}(S^2) = \frac{2\sigma^4}{n-1}.$$

This last result explains why we used the factor $(n-1)^{-1}$ instead of n^{-1} in the definition of S^2. This definition gives us a statistic whose expectation is equal to the true variance. Finally, we mention here a result that we will find quite useful later.

Theorem 2.9 (Student's Statistic and Its Sampling Distribution) *Let* $X_1, \ldots, X_n \overset{iid}{\sim} N(\mu, \sigma^2)$. *Then,*

$$\frac{\bar{X} - \mu}{S/\sqrt{n}} \sim t_{n-1}.$$

Here t_{n-1} denotes Student's distribution with $n-1$ degrees of freedom.

Definition 2.10 (Student's t Distribution)

A random variable X is said to follow Student's t distribution with parameter $k \in \mathbb{N}$ (called the number of degrees of freedom), denoted $X \sim t_k$, if,

$$f_X(x; k) = \frac{\Gamma\left(\frac{k+1}{2}\right)}{\Gamma\left(\frac{k}{2}\right)\sqrt{k\pi}} \left(1 + \frac{x^2}{k}\right)^{-\frac{k+1}{2}},$$

Assuming $k > 2$, the mean and variance of $X \sim t_k$ are given by

$$\mathbb{E}[X] = 0, \qquad \mathrm{Var}[X] = \frac{k}{k-2}.$$

The mean is undefined for $k = 1$ and the variance is undefined for $k \leq 2$. The moment generating function is undefined for any $k \in \mathbb{N}$.

Proof of Theorem 2.9 Let $Z = (\bar{X} - \mu)/(\sigma/\sqrt{n})$ and $V = (n-1)S^2/\sigma^2$, and note that

$$T = \frac{Z}{\sqrt{\frac{V}{n-1}}} = \frac{\bar{X} - \mu}{S\sqrt{n}}.$$

Thus, to prove the theorem, we will find the density of T. To this aim, we observe that by Proposition 2.7 (p. 45) we have

1. Z is a standard normal random variable.
2. V is a χ^2_{n-1} random variable.
3. Z and V are independent.

We will first find the joint density of (T, V), and then integrate to find the marginal of T. To this aim, consider the transformation

$$g : (Z, V) \mapsto (T, V) = \left(\frac{Z}{\sqrt{V/(n-1)}}, V \right)$$

whose inverse is given by

$$g^{-1} : (T, V) \mapsto \left(T\sqrt{\frac{V}{n-1}}, V \right)$$

and has a corresponding upper triangular Jacobian

$$J_{g^{-1}} = \begin{pmatrix} \sqrt{V/(n-1)} & T\frac{V^{-1/2}}{2\sqrt{(n-1)}} \\ 0 & 1 \end{pmatrix} \quad \Rightarrow \quad \det(J_{g^{-1}}(t, v)) = \sqrt{\frac{v}{n-1}}.$$

Since Z and V are independent, it follows that

$$f_{Z,V}(z, v) = f_Z(z) f_V(v) = \frac{1}{2^{\frac{n}{2}} \pi^{\frac{1}{2}} \Gamma\left(\frac{n-1}{2}\right)} v^{\frac{n-1}{2}-1} e^{-\frac{1}{2}(v+z^2)}.$$

The joint density of (T, V) is thus given by

$$\begin{aligned} f_{T,V}(t, v) &= f_{Z,V}(g^{-1}(t, v)) |\det(J_{g^{-1}}(t, v))| \\ &= \frac{1}{2^{\frac{n}{2}} \pi^{\frac{1}{2}} \Gamma\left(\frac{n-1}{2}\right)} v^{\frac{n-1}{2}-1} e^{-\frac{1}{2}(v+v\frac{t^2}{n-1})} \cdot \left(\frac{v}{n-1}\right)^{\frac{1}{2}} \\ &= \frac{1}{2^{\frac{n}{2}} \sqrt{\pi(n-1)} \Gamma\left(\frac{n-1}{2}\right)} \cdot v^{\frac{n-2}{2}} e^{-\frac{v}{2}(1+\frac{t^2}{n-1})}. \end{aligned}$$

It now remains to integrate out v, and find the marginal density of T:

$$f_T(t) = \frac{1}{2^{\frac{n}{2}} \Gamma\left(\frac{n-1}{2}\right) \sqrt{(n-1)\pi}} \int e^{-\frac{v}{2}\left(\frac{t^2}{n-1}+1\right)} v^{\frac{n-2}{2}} dv.$$

Putting

$$y = \frac{v}{2}\left(\frac{t^2}{n-1} + 1\right),$$

we obtain

$$v = \frac{2y}{\left(\frac{t^2}{n-1}+1\right)} \quad \text{and} \quad dv = \frac{2}{\left(\frac{t^2}{n-1}+1\right)},$$

and thus

$$f_T(t) = \frac{1}{2^{\frac{n}{2}}\Gamma\left(\frac{n-1}{2}\right)\sqrt{(n-1)\pi}} \cdot \int e^{-y} \cdot \left[(2y)\left(\frac{t^2}{n-1}+1\right)^{-1}\right]^{\frac{n-2}{2}} \cdot 2\left(\frac{t^2}{n-1}+1\right)^{-1} dy$$

$$= \frac{1}{2^{\frac{n}{2}}\Gamma\left(\frac{n-1}{2}\right)\sqrt{(n-1)\pi}} \cdot \left(\frac{t^2}{n-1}+1\right)^{-\frac{n}{2}} \cdot 2^{\frac{n}{2}} \cdot \int y^{\frac{n-2}{2}} e^{-y} dy$$

$$= \frac{\Gamma\left(\frac{n}{2}\right)}{\Gamma\left(\frac{n-1}{2}\right)} \cdot \frac{1}{\sqrt{(n-1)\pi}} \cdot \left(\frac{t^2}{n-1}+1\right)^{-\frac{n}{2}} \cdot \int \frac{1}{\Gamma\left(\frac{n}{2}\right)} \cdot y^{\frac{n}{2}-1} e^{-y} dy$$

$$= \frac{\Gamma\left(\frac{n}{2}\right)}{\Gamma\left(\frac{n-1}{2}\right)} \cdot \frac{1}{\sqrt{(n-1)\pi}} \cdot \left(\frac{t^2}{n-1}+1\right)^{-\frac{n}{2}}.$$

where the integral in the penultimate line is equal to 1, being the integral of a $\Gamma(n/2, 1)$ density function.   □

2.3 Sampling from an Exponential Family

In the previous paragraph we were able to determine the joint distribution of a normal random sample $X_1, \ldots, X_n$, the sampling distribution of two key statistics, and the moments of these two key statistics. What if the distribution we are sampling from is not normal, but binomial, or Poisson, or exponential? More generally: what if the sample $X_1, \ldots, X_n$ comes from some other exponential family? In other words, let $X_1, \ldots, X_n \overset{iid}{\sim} f$, where

$$f(x) = \exp\left\{\sum_{i=1}^{k} \phi_i T_i(x) - \gamma(\phi_1, \ldots, \phi_k) + S(x)\right\}, \qquad x \in \mathcal{X}.$$

1. Is it possible to find the joint distribution of a sample $(X_1, \ldots, X_n)$?
2. Is it possible to find the exact moments of some key statistics?
3. Is it possible to find the exact sampling distribution of some important statistics?
   The next theorem gives an affirmative answer to the first two questions. Unfortunately, the answer to the last question is: it's complicated. For simplicity, we will

focus on 1-parameter exponential families, but the results can easily be suitably generalised to the k-parameter case.

Proposition 2.11 (Sampling from an Exponential Family) *Let* $X_1, \ldots, X_n \overset{iid}{\sim}$ f, *where*

$$f(x) = \exp\{\phi T(x) - \gamma(\phi) + S(x)\}, \qquad x \in \mathcal{X}$$

where $\phi \in \Phi \subseteq \mathbb{R}$, *be a density of a 1-parameter exponential family form. Then:*

1. *The joint density of* $(X_1, \ldots, X_n)$ *is of a 1-parameter exponential family form, given by*

$$f_{X_1, \ldots, X_n}(x_1, \ldots, x_n) = \exp\left\{\phi\tau(x_1, \ldots, x_n) - n\gamma(\phi) + \sum_{i=1}^{n} S(x_i)\right\}, \qquad x_i \in \mathcal{X},$$

where

$$\tau(x_1, \ldots, x_n) = \sum_{i=1}^{n} T(x_i).$$

2. *If* Φ *is open, then* γ *is infinitely differentiable, and*

$$\mathbb{E}[\tau(X_1, \ldots, X_n)] = n\gamma'(\phi) < \infty \quad \text{and} \quad \text{Var}[\tau(X_1, \ldots, X_n)] = n\gamma''(\phi) < \infty.$$

▶ **Remark 2.12** The theorem demonstrates why τ is a key statistic that we are interested in: by the Fisher–Neyman factorisation theorem we can immediately see that τ is sufficient for ϕ (if $\phi = \eta(\theta)$ for some 1-1 mapping $\eta(\cdot)$, then it is clear that τ is also sufficient for θ).

▶ **Remark 2.13** The sampling distribution of the sufficient statistic τ is still of a 1-parameter exponential family form, with the same natural parameter ϕ and with the identity as a natural statistic, i.e. it is of the form

$$f_\tau(t) = \exp\{\phi t - A(\phi) + B(t)\},$$

for some $A : \Phi \to \mathbb{R}$ and $B : \mathbb{R} \to \mathbb{R}$ (we will not prove this because it requires measure theory). However, an explicit general form of the density cannot be given (i.e. we cannot find a general formula for the form of the functions A and B). For a simple general formula, we will need to resort to approximations of this sampling distribution, and this we do in the next section. Nevertheless, we can indeed determine general formulae for the mean and variance of $\tau(X_1, \ldots, X_n)$.

▶ **Remark 2.14** The fact that γ is infinitely differentiable when Φ is open (conclusion 2 of the Proposition) will be taken for granted for the remainder of the text.

Proof of Proposition (2.11) Part (1) is immediate from independence and from the form of a 1-parameter exponential family. To prove (2), we first calculate the MGF of $T(X_i)$, some $i \leq n$.

$$M_T(u) = \int_{\mathcal{X}} \exp\{uT(x)\} \exp\{\phi T(x) - \gamma(\phi) + S(x)\} dx$$

$$= \exp\{\gamma(u + \phi) - \gamma(\phi)\} \int_{\mathcal{X}} \exp\{(u + \phi)T(x) - \gamma(u + \phi) + S(x)\} dx.$$

Since Φ is open, there exists an ϵ such that $(u + \phi) \in \Phi$ if $|u| < \epsilon$. Thus $u + \phi$ is a valid parameter when $|u| < \epsilon$, yielding $\int_{\mathcal{X}} \exp\{(u+\phi)T(x) - \gamma(u+\phi) + S(x)\} dx = 1$. We conclude:

$$M_T(u) = \exp\{\gamma(u + \phi) - \gamma(\phi)\}, \qquad |u| < \epsilon. \tag{2.1}$$

Since the moment generating function exists for $|u| < \epsilon$, it follows from Proposition A.8 (p. 163) that M_T is infinitely differentiable for $|u| < \epsilon$ and so it also must be that γ is infinitely differentiable on Φ. Furthermore, Proposition A.8 (p. 163) also implies that all moments of $T(X_i)$ exist, for all values of $\phi \in \Phi$; and

$$\mathbb{E}[T(X_i)] = \left. \frac{d}{du} M_T(u) \right|_{u=0} = \gamma'(\phi)$$

$$\mathbb{E}[T^2(X_i)] = \left. \frac{d^2}{du^2} M_T(u) \right|_{u=0} = \gamma''(\phi) + [\gamma'(\phi)]^2.$$

We conclude that $\mathbb{E}[T(X_i)] = \gamma'(\phi)$ and that $\mathrm{Var}[T(X_i)] = \mathbb{E}[T^2(X_i)] - \mathbb{E}^2[T(X_i)] = \gamma''(\phi)$. It now immediately follows by independence of $X_1, \ldots, X_n$ that

$$\mathbb{E}[\tau(X_1, \ldots, X_n)] = \mathbb{E}\left[\sum_{i=1}^{n} T(X_i) \right] = \sum_{i=1}^{n} \mathbb{E}[T(X_i)] = n\gamma'(\phi)$$

$$\mathrm{Var}[\tau(X_1, \ldots, X_n)] = \mathrm{Var}\left[\sum_{i=1}^{n} T(X_i) \right] = \sum_{i=1}^{n} \mathrm{Var}[T(X_i)] = n\gamma''(\phi).$$

$\square$

Exercise 23 Let $X_1, \ldots, X_n \overset{iid}{\sim} f$, where f is of an exponential family form, expressed in the usual parametrisation as $f(x) = \exp\left[\eta(\theta)T(x) - d(\theta) + S(x)\right]$. Assuming that Θ is open, show that:

1. If η is k-times continuously differentiable ($k \geq 1$), invertible, and $\eta'(\theta) \neq 0$, then d is k-times continuously differentiable.
2. If η is twice continuously differentiable and invertible with $\eta'(\theta) \neq 0$, then

$$\mathbb{E}[\tau(X_1, \ldots, X_n)] = n\frac{d'(\theta)}{\eta'(\theta)} \quad \& \quad \mathrm{Var}[\tau(X_1, \ldots, X_n)] = n\frac{d''(\theta)\eta'(\theta) - d'(\theta)\eta''(\theta)}{[\eta'(\theta)]^3}$$

Hint: use the inverse function theorem (Theorem A.2, p. 159).

▶ **Remark 2.15** The fact that d is k-times continuously differentiable (for $k \geq 1$) whenever Θ is open and η is k-times continuously differentiable, invertible, and has a non-vanishing derivative (see part (i) of the exercise) will be taken for granted in the rest of the text without special mention.

2.4 Approximate Sampling Distributions

We saw in the last section that the sampling distribution of the sufficient statistic $\tau(X_1, \ldots, X_n)$ when sampling from a one-parameter exponential may not be straightforward to determine exactly. For this reason, we will often try to approximate it, assuming that the sample size n is large enough. This requires a mathematical notion of what it means to say that the distribution $F_{\tau(X_1, \ldots, X_n)}$ is approximately given by some other distribution G. If we see $F_{\tau(X_1, \ldots, X_n)}$ as a sequence of distribution functions F_n indexed by the sample size n, then approximation by G should be formalised by some notion of convergence of F_n to G as $n \to \infty$. The appropriate type of convergence is called *convergence in distribution*.

Definition 2.16 (Convergence in Distribution)

Let $\{F_n\}_{n \geq 1}$ be a sequence of distribution functions and G be a distribution function on $\mathbb{R}$. We say that F_n converges in distribution to G, and write $F_n \overset{d}{\longrightarrow} G$, if and only if

$$F_n(x) \overset{n \to \infty}{\longrightarrow} G(x),$$

for all x that are continuity points of G.

▶ **Remark 2.17** Notice that convergence in distribution is similar to pointwise convergence of the sequence of distributions, except that we do not insist to have pointwise convergence at the discontinuity points of the limit (recall that any distribution is cadlag: continuous from the right, and has limits from the left).

Example 2.18 (Maximum of Uniform Random Variables)

Let $X_1, \ldots, X_n \overset{iid}{\sim} Unif(0, 1)$, $M_n = \max\{X_1, \ldots, X_n\}$, and $Q_n = n(1 - M_n)$.

$$\mathbb{P}[Q_n \leq x] = \mathbb{P}[M_n \geq 1 - x/n] = 1 - \left(1 - \frac{x}{n}\right)^n \overset{n \to \infty}{\longrightarrow} 1 - e^{-x}.$$

Note that the limit is the distribution function of an $Exp(1)$ random variable. □

Exercise 24 (Law of Rare Events) Let $\{X_n\}_{n \geq 1}$ be a sequence of $Binom(n, p_n)$ random variables, such that $p_n = \lambda/n$, for some constant $\lambda > 0$. Prove that $X_n \overset{d}{\longrightarrow} Y$, where $Y \sim Poisson(\lambda)$.

When $F_n(x) = \mathbb{P}[X_n \leq x]$ for some sequence of random variables $\{X_n\}_{n \geq 1}$ and $G(x) = \mathbb{P}[Z \leq x]$ for some other random variable Z, we will abuse notation and write

$$X_n \overset{d}{\longrightarrow} Z.$$

This will be taken to mean that the distribution of X_n can be approximated, for large n, by the distribution of Z. So, if we denote $\tau_n = \tau(X_1, \ldots, X_n)$, then the problem of determining an approximate distribution for $\tau(X_1, \ldots, X_n)$ is equivalent to the problem of finding some random variable Z whose distribution is explicitly known, and such that $\tau_n \overset{d}{\longrightarrow} Z$. We will give a partial solution to this problem in the next two subsections.

Before we conclude this introduction, we introduce a second type of convergence that merits independent consideration.

Definition 2.19 (Convergence in Probability)

If a sequence of random variables $\{X_n\}$ is such that $\mathbb{P}[|X_n - Y| > \epsilon] \overset{n \to \infty}{\longrightarrow} 0$ for all $\epsilon > 0$ and for some other random variable Y, we say that X_n converges in probability to Y and write $X_n \overset{p}{\longrightarrow} Y$.

In general $X_n \overset{p}{\longrightarrow} Y \implies X_n \overset{d}{\longrightarrow} Y$, but the converse may fail to hold true:

Exercise 25 Let $\{X_n\}_{n=1}^{\infty}$ be a sequence of random variables with

$$X_n = (-1)^n X, \quad \mathbb{P}[X = -1] = \mathbb{P}[X = 1] = \frac{1}{2}.$$

Show that $X_n \overset{d}{\to} X$, but $X_n \overset{p}{\nrightarrow} X$.

Suppose, though, that $Y = c \in \mathbb{R}$ is a constant, and $\{X_n\}_{n \geq 1}$ is a sequence such that $X_n \xrightarrow{d} c$. Then we have the following result.

Lemma 2.20 *Let $\{X_n\}_{n \geq 1}$ be a sequence of random variables taking values in $\mathbb{R}$, and $c \in \mathbb{R}$ be some constant. Then,*

$$X_n \xrightarrow{d} c \iff \mathbb{P}[|X_n - c| > \epsilon] \xrightarrow{n \to \infty} 0, \quad \forall \epsilon > 0.$$

Exercise 26 Prove the last lemma.

2.4.1 Approximate Distributions for Sums

It was seen in Proposition 2.11 (p. 51) that the sufficient statistic for an iid sample $X_1, \ldots, X_n$ from a one-parameter exponential family

$$f(x) = \exp\{\phi T(x) - \gamma(\phi) + S(x)\}$$

is of the form $\tau(X_1, \ldots, X_n) = \sum_{i=1}^{n} T(X_i)$, where

$$\mathbb{E}[\tau(X_1, \ldots, X_n)] = n\gamma'(\phi) < \infty \quad \text{and} \quad \text{Var}[\tau(X_1, \ldots, X_n)] = n\gamma''(\phi) < \infty.$$

If we define

$$\overline{T}_n = \frac{1}{n}\tau(X_1, \ldots, X_n) = \frac{1}{n}\sum_{i=1}^{n} T(X_i),$$

then we notice that we have a random variable that is built as the average of n iid random variables, of finite mean $\gamma'(\phi)$ and finite variance $\gamma''(\phi)/n$. Though the exact sampling behaviour of such averages might not always be tractable, their behaviour for large n becomes surprisingly simple. The goal of this section is to describe this behaviour. In other words, given $Y_1, \ldots, Y_n$ iid random variables with $\mathbb{E}[Y_i] = \mu < \infty$ and $\text{Var}[Y_i] = \sigma^2 < \infty$, we wish to study the approximate distribution of $\sum_{i=1}^{n} Y_t$.

We note that the expectation of $\sum_{i=1}^{n} Y_i$ is $n\mu$, which tends to infinity as n grows. Therefore, we cannot hope to get a distributional approximation if we do not tame this explosion. The first idea that comes to mind is to simply divide by n. That is, to look at the empirical mean $\overline{Y}_n = \frac{1}{n}\sum_{i=1}^{n} Y_i$ instead. The expectation of this empirical mean is μ, which remains constant with respect to n. By Chebyshev's inequality (Lemma A.4, p. 159), we have that

$$\mathbb{P}[|\overline{Y}_n - \mu| > \epsilon] \leq \frac{\sigma^2}{n\epsilon^2} \xrightarrow{n \to \infty} 0, \quad \forall \epsilon > 0.$$

Theorem 2.21 (L^2 Weak Law of Large Numbers) *Let $Y_1, \ldots, Y_n$ be iid random variables such that $\mathbb{E}[Y_i] = \mu < \infty$ and $\mathrm{Var}[Y_i] = \sigma^2 < \infty$. Let $\overline{Y}_n = \frac{1}{n} \sum_{i=1}^{n} Y_i$. Then,*

$$\overline{Y}_n \xrightarrow{p} \mu.$$

▶ **Remark 2.22 (L^1 Weak Law of Large Numbers)** Actually, the same conclusion can be drawn under weaker assumptions: it suffices to assume that $\mathbb{E}|Y_i| < \infty$, rather than $\mathrm{Var}[Y_i] < \infty$.

Consequently, the realisations of the random variable $\overline{Y}_n$ become more and more concentrated around its mean as n grows, i.e. $(\overline{Y} - \mu) \xrightarrow{p} 0$. How does $\overline{Y}_n$ vary around μ, though, as n grows? The factor n^{-1} was such that it made $n^{-1} \sum_{i=1}^{n} Y_i$ converge to a constant. The reason is that multiplying with the factor n^{-1} made the variance equal to σ^2/n, and hence made it converge to zero. The key observation is that the mean of $c \times \sum_{i=1}^{n} Y_i$ scales linearly in c but its variance scales quadratically in c. To get a finer approximation we need to consider the re-scaled differences $\sqrt{n}(\overline{Y} - \mu)$. Notice that these have variance σ^2 for all n. The following remarkable result tells us that these scaled differences are approximately normal:

Theorem 2.23 (Central Limit Theorem) *Let $Y_1, \ldots, Y_n$ be iid random variables such that $\mathbb{E}[Y_i] = \mu < \infty$ and $\mathrm{Var}[Y_i] = \sigma^2 < \infty$. Let $\overline{Y}_n = \frac{1}{n} \sum_{i=1}^{n} Y_i$. Then,*

$$\sqrt{n}(\overline{Y}_n - \mu) \xrightarrow{d} N(0, \sigma^2).$$

We discuss the proof of the Central Limit Theorem in Sect. A.8 (p. 173). We now have an immediate corollary, by combining the Central Limit Theorem with Proposition 2.11 (p. 51), that will be very useful for statistical inference:

Corollary 2.24 (Approximate Sampling Distribution in Exponential Families) *Let $X_1, \ldots, X_n \overset{iid}{\sim} f$, where*

$$f(x) = \exp\{\phi T(x) - \gamma(\phi) + S(x)\}, \qquad x \in \mathcal{X}$$

where $\phi \in \Phi \subseteq \mathbb{R}$. Let

$$\overline{T}_n = \frac{1}{n} \sum_{i=1}^{n} T(X_i) = n^{-1} \tau(X_1, \ldots, X_n).$$

If Φ is open, then

$$\sqrt{n}(\overline{T}_n - \gamma'(\phi)) \xrightarrow{d} N(0, \gamma''(\phi)).$$

2.4.2   Approximate Distributions for Functions of Sums

What if the statistic whose sampling distribution we wish to determine is not just a sum of iid random variables, but some smooth function of a sum? For example, suppose that we wish to consider a statistic of the form $g(\overline{Y}_n)$ rather than $\overline{Y}_n$ itself. Can we say anything about the asymptotic behaviour of this new random variable? The next three results give us affirmative answers to this question in some important special cases.

Theorem 2.25 (Continuous Mapping Theorem) *If X is a random variable such that $\mathbb{P}[X \in \mathcal{A}] = 1$ and $g : \mathbb{R} \to \mathbb{R}$ is continuous everywhere on $\mathcal{A}$, then*

$$X_n \xrightarrow{d} X \implies g(X_n) \xrightarrow{d} g(X).$$

Proof See Sect. A.7 (p. 169)   □

Theorem 2.26 (Slutsky's Theorem) *Let X be a random variable and $c \in \mathbb{R}$ a constant. If $X_n \xrightarrow{d} X$ and $Y_n \xrightarrow{p} c \in \mathbb{R}$, then it follows that $X_n + Y_n \xrightarrow{d} X + c$ and $X_n Y_n \xrightarrow{d} cX$, as $n \to \infty$.*

Proof See Sect. A.7 (p. 169)   □

It's important to note that, in general, one cannot replace the constant $c \in \mathbb{R}$ with a non-degenerate random variable, say Y, in Slutsky's theorem. The problem is that we have no information on the joint distribution of (X_n, Y_n). For a simple counterexample, take $X_n = -Z + n^{-1}$ and $Y_n = Z - n^{-1} = -X_n$, for $Z \sim N(0, 1)$. Then, $X_n \xrightarrow{d} Z$ (since $-Z \sim N(0, 1)$), $Y_n \xrightarrow{p} Z$, but for all n, we have $X_n + Y_n = 0$, and thus $X_n + Y_n$ fails to converge in distribution to $2Z$.

Theorem 2.27 (The Delta Method) *Let $Z_n := a_n(X_n - \theta) \xrightarrow{d} Z$ where $a_n, \theta \in \mathbb{R}$ for all n and $a_n \uparrow \infty$. Let $g : \mathbb{R} \to \mathbb{R}$ be differentiable at θ. Then, $a_n(g(X_n) - g(\theta)) \xrightarrow{d} g'(\theta)Z$, provided that $g'(\theta) \neq 0$.*

Proof Taylor expanding (Theorem A.1, p. 159) around θ gives

$$g(X_n) = g(\theta) + g'(\theta_n^*)(X_n - \theta),$$

where θ_n^* lies between X_n and θ. Thus $|\theta_n^* - \theta| < |X_n - \theta| = a_n^{-1} \cdot |a_n(X_n - \theta)| = a_n^{-1} Z_n \xrightarrow{p} 0$ as a result of Slutsky's theorem. Therefore, $\theta_n^* \xrightarrow{p} \theta$. By the continuous mapping theorem it now follows that $g'(\theta_n^*) \xrightarrow{p} g'(\theta)$. Consequently,

$$a_n(g(X_n) - g(\theta)) = a_n(g(\theta) + g'(\theta_n^*)(X_n - \theta) - g(\theta))$$

$$= g'(\theta_n^*)a_n(X - \theta) \xrightarrow{d} g'(\theta)Z,$$

using Slutsky's theorem once again.   □

These three results enable us to obtain new limit theorems (new approximations) from old ones. For example, the central limit theorem tells us that if $Y_1, \ldots, Y_n$ are iid with mean μ and finite variance $\sigma^2 < \infty$, then $\sqrt{n}(\bar{Y}_n - \mu) \xrightarrow{d} N(0, \sigma^2)$. Now, the delta method implies that

$$\sqrt{n}(g(\bar{Y}_n) - g(\mu)) \xrightarrow{d} N(0, \sigma^2(g'(\mu))^2),$$

for all continuously differentiable functions g. Now let W_n be a sequence of random variables such that $W_n \xrightarrow{p} \sigma$. It is an easy exercise to use Slutksy's theorem and conclude that

$$\sqrt{n}\left(\frac{g(\bar{Y}_n) - g(\mu)}{W_n}\right) \xrightarrow{d} N(0, (g'(\mu))^2).$$

Exercise 27 Let $X_1, \ldots, X_n \overset{iid}{\sim} Pois(\lambda)$, where $\lambda \in (0, \infty)\backslash\{1\}$ and consider the probability $\pi = \mathbb{P}(X_i = 1) = \lambda e^{-\lambda}$. We wish to approximate π by $\hat{\pi}_n = \hat{\lambda}_n e^{-\hat{\lambda}_n}$ where $\hat{\lambda}_n = \frac{1}{n}\sum_{i=1}^{n} X_i$ (effectively replacing the true mean in the expression for the probability by the empirical mean). We know that the $\hat{\lambda}$ satisfies the central limit theorem. Show that this also gives a central limit theorem for $\hat{\pi}_n$ in the form of

$$\frac{\sqrt{n}(\hat{\pi}_n - \pi)}{\sqrt{\hat{\lambda}_n e^{-\hat{\lambda}_n}(1 - \hat{\lambda}_n)}} \xrightarrow{d} Y,$$

where $Y \sim N(0, 1)$. Hint: you will need to use the central limit theorem, the delta method, the law of large number, and Slutsky's theorem.

Exercise 28 Let $x_1, \ldots, x_n$ be independent realisations of a random variable X possessing a continuous density function f. Show that the histogram $hist_{x_1, \ldots, x_n}(y)$ converges in probability pointwise to $f(y)$, as $n \to \infty$, $h_n \to 0$ and $nh_n \to \infty$. Hint: the number of observations in the interval I_{j_n}, given by $N_n = \sum_{i=1}^{n} 1_{\{x_i \in I_{j_n}\}}$, follows a *Binom*$(n, p_n)$ distribution, where $p_n = \int_{I_{j_n}} f(x)dx$. You will need to use the fact that

$$\left| \frac{N_n}{nh_n} - f(y) \right| \leq \left| \frac{N_n}{nh_n} - \frac{p_n}{h_n} \right| + \left| \frac{p_n}{h_n} - f(y) \right|,$$

as well as Chebyshev's inequality (Lemma A.4, p. 159).

Point Estimation of Model Parameters

3

We now return to the bigger picture: we are modelling a stochastic phenomenon by a regular parametric family of distributions $\mathcal{F} = \{F_\theta : \theta \in \Theta\}$, where $\Theta \subseteq \mathbb{R}^p$. We observe n independent and identically distributed outcomes from the phenomenon, say $X_1, \ldots, X_n \overset{iid}{\sim} F_\theta$ for some $\theta_0 \in \Theta$, but do not know/observe the $\theta \in \Theta$ that generated them (the *true state of nature*). With this iid sample at our disposal, we wish to make inferences about θ. Perhaps the most obvious inference we may wish to draw is: which is the θ that generated the sample $X_1, \ldots, X_n$? This is known as the problem of *point estimation*. Since $X_1, .., X_n$ is all we have available to estimate the value of θ, we will use some function of the sample as an estimator.

Definition 3.1 (Point Estimator)

A statistic whose range is contained in Θ is called a *point estimator*. Equivalently, a point estimator is a statistic $T : \mathcal{X}^n \to \Theta$.

▶ **Remark 3.2** Since the purpose of an estimator is to provide a guess of the true θ that generated the data, we typically denote it by $\hat{\theta}$. Note that θ is a deterministic parameter, but $\hat{\theta}$ is a random variable, since $\hat{\theta} = T(X_1, \ldots, X_n)$.

Clearly, the purpose of an estimator is to estimate the unknown parameter. But, according to the definition, essentially any function of the sample that maps into Θ could be an estimator. Which one should we pick? Or, even more simply, if we are presented with an estimator $\hat{\theta}$ how can we judge its quality?

The important thing here is that estimators are *random variables*. Therefore, for every realisation of the sample $X_1, \ldots, X_n$ the estimator $\hat{\theta}$ will take a different value. A good estimator should be such that its typical realisations fall "close" to θ. In other words, the distribution of a good estimator is concentrated around the value of the true parameter θ.

© Springer International Publishing Switzerland 2016
V.M. Panaretos, *Statistics for Mathematicians*, Compact Textbooks in Mathematics,
DOI 10.1007/978-3-319-28341-8_3

3.1   Criteria for Comparing Estimators

Still however, the question remains: how can we measure the *concentration of the distribution of* $\hat{\theta}$? There are many different criteria that one could use, but there are two basic concentration characteristics that statisticians typically focus on: the mean and the variance of $\hat{\theta}$. *Why?*

1. One reason is that the mean and the variance are easy to interpret: the mean $\mathbb{E}[\hat{\theta}]$ tells us how close to the target our estimator is on average. And $\mathrm{Var}[\hat{\theta}]$ tells us how dispersed our estimator is around its average. If both are small, we should have reasonable concentration.
2. A second reason is that the exact distribution of $\hat{\theta}$ is often unknown. As we saw in previous sections, we then need to resort to asymptotic approximations. Relatively often, it happens that the approximate distribution of $\hat{\theta}$ is normal. And, for the normal distribution, the mean and the variance capture all of the concentration characteristics.
3. Even if the approximate distribution is not normal, concentration inequalities such as Markov's inequality or Chebyschev's inequality (Lemmas A.3, p. 159 and A.4, p. 159) can be used to bound the probability $\mathbb{P}\{\|\hat{\theta} - \theta\| > \epsilon\}$ (which expresses concentration) given knowledge of a mean and a variance. Such inequalities are valid regardless of the precise distribution of $\hat{\theta}$.

   It turns out that the so-called *mean squared error* takes both the mean and the variance into account.

Definition 3.3 (Mean Squared Error)

Let $\hat{\theta}$ be an estimator for a parameter θ of a parametric model $\{F_\theta : \theta \in \Theta\}$. The mean squared error of $\hat{\theta}$ is defined to be

$$MSE(\hat{\theta}, \theta) = \mathbb{E}[\|\hat{\theta} - \theta\|^2].$$

Notice that the MSE depends on both our estimator and the true state of nature. Therefore, an estimator $\hat{\theta}$ may perform well if the true θ is in some region of the parameter space Θ, but not as well in other regions of the parameter space. We will revisit this issue later.

For the moment, though, we see why the MSE is connected with the mean and the variance of $\hat{\theta}$:

Lemma 3.4 (Bias-Variance Decomposition) *Write* $\hat{\theta} = (\hat{\theta}_1, \ldots, \hat{\theta}_p)^\top$. *The mean squared error of an estimator admits the decomposition*

$$MSE(\hat{\theta}, \theta) = \|\mathbb{E}[\hat{\theta}] - \theta\|^2 + \mathbb{E}[\|\hat{\theta} - \mathbb{E}(\hat{\theta})\|^2] = \|bias(\hat{\theta}, \theta)\|^2 + \sum_{k=1}^{p} \mathrm{Var}[\hat{\theta}_k].$$

▶ **Remark 3.5** We call the quantity $\mathbb{E}[\hat{\theta} - \theta] = bias(\hat{\theta}, \theta)$ the bias of the estimator $\hat{\theta}$ at true parameter θ. It expresses how far off $\hat{\theta}$ is from θ on average. When the bias at some coordinate of θ is positive we have *overestimation*; when it is negative we have *underestimation*; when the bias is zero, we speak of an *unbiased estimator*. Notice that the variances $\mathrm{Var}[\hat{\theta}_k]$ can also depend on θ, even though this is not explicitly reflected in the notation.

Proof of Lemma 3.4 We expand the MSE after adding and subtracting $\mathbb{E}[\hat{\theta}]$:

$$\mathbb{E}[\|\hat{\theta} - \theta\|^2] = \mathbb{E}[\|\hat{\theta} - \mathbb{E}[\hat{\theta}] + \mathbb{E}[\hat{\theta}] - \theta\|^2]$$

$$= \mathbb{E}\left[(\hat{\theta} - \mathbb{E}[\hat{\theta}] + \mathbb{E}[\hat{\theta}] - \theta)^\top (\hat{\theta} - \mathbb{E}[\hat{\theta}] + \mathbb{E}[\hat{\theta}] - \theta)\right]$$

$$= \|\mathbb{E}[\hat{\theta}] - \theta\|^2 + \mathbb{E}[\|\hat{\theta} - \mathbb{E}(\hat{\theta})\|^2] + 2\mathbb{E}\left[(\hat{\theta} - \mathbb{E}[\hat{\theta}])^\top (\mathbb{E}[\hat{\theta}] - \theta)\right]$$

$$= \|\mathbb{E}[\hat{\theta}] - \theta\|^2 + \mathbb{E}[\|\hat{\theta} - \mathbb{E}(\hat{\theta})\|^2] + 2\underbrace{(\mathbb{E}[\hat{\theta}] - \mathbb{E}[\hat{\theta}])}_{=0}^\top (\mathbb{E}[\hat{\theta}] - \theta)$$

$$= \|\mathbb{E}[\hat{\theta}] - \theta\|^2 + \sum_{k=1}^{p} \mathbb{E}[(\hat{\theta}_k - \mathbb{E}(\hat{\theta}_k))^2],$$

by linearity of the expectation and since $(\mathbb{E}[\hat{\theta}] - \theta)$ is deterministic. $\square$

Exercise 29 (Unbiased Estimators Don't Always Exist) Let $Y \sim Binom(n, p)$, where $p \in (0, 1)$.
1. Show that Y/n is an unbiased estimator of p.
2. Show that there exists no unbiased estimator of $1/p$.
3. Show that there exists no unbiased estimator of the natural parameter $\phi = \log\left(\frac{p}{1-p}\right)$.
Remark: ϕ is called the *log odds ratio*.

As was noted earlier, the concentration of an estimator $\hat{\theta}$ around the true parameter θ can always be bounded using the mean squared error (provided that the estimator $\hat{\theta}$ has finite variance).

Lemma 3.6 *Let $\hat{\theta}$ be an estimator of $\theta \in \mathbb{R}^p$ such that $\mathrm{Var}[\hat{\theta}] < \infty$. Then, for all $\epsilon > 0$,*

$$\mathbb{P}[\|\hat{\theta} - \theta\| > \epsilon] \le \frac{MSE(\hat{\theta}, \theta)}{\epsilon^2}.$$

Proof Let $X = \|\hat{\theta} - \theta\|^2$. Since $\epsilon > 0$, Markov's inequality (Lemma A.3, p. 159) yields

$$\mathbb{P}[\|\hat{\theta} - \theta\| > \epsilon] = \mathbb{P}[X > \epsilon^2] \leq \frac{E[X]}{\epsilon^2} = \frac{\mathbb{E}[\|\hat{\theta} - \theta\|^2]}{\epsilon^2} = \frac{MSE(\hat{\theta}, \theta)}{\epsilon^2}.$$

$\square$

Let $\hat{\theta}_n = T(X_1, \ldots, X_n)$ be an estimator of a parameter θ (we write the subscript n to emphasize the dependence on the sample size). Notice that if $MSE(\hat{\theta}_n, \theta)$ converges to zero as $n \to \infty$, then the previous result implies that $\hat{\theta}_n \xrightarrow{p} \theta$. When an estimator has this last property, we call the estimator consistent.

Definition 3.7 (Consistency)

An estimator $\hat{\theta}_n$ of θ constructed on the basis of a sample of size n is called consistent if $\hat{\theta}_n \xrightarrow{p} \theta$ as $n \to \infty$.

▶ **Remark 3.8** Notice that convergence of the MSE to zero implies consistency. The converse is not true in general, though.

Though we will focus on the mean squared error, it is certainly not the only criterion for judging the performance of an estimator: there are many other criteria that can be imagined. In general, one can define a *loss function*, $\mathcal{L} : \Theta \times \Theta \to [0, \infty)$, which represents the loss incurred when we estimate θ by $\hat{\theta}$. Then, one uses the average loss, or *risk*, as a measure of performance: $R(\hat{\theta}, \theta) = \mathbb{E}[\mathcal{L}(\hat{\theta}, \theta)]$. The "goodness" or "badness" of an estimator will clearly depend on our choice of loss function, and so this choice must be made judiciously. Notice that the mean squared error is the risk function obtained when the loss function is defined to be the squared Euclidean distance.

3.2   Fundamental Limitations to Estimation Accuracy

We can use the mean squared error to compare any two candidate estimators, and so have an idea of their relative performance. It would also be nice to have a more absolute benchmark in order to compare the mean square error of any single estimator to a *best achievable* mean square error for the given problem. It turns out that this is a difficult problem, because it is equivalent to finding a uniformly optimal estimator: an estimator T_* such that $MSE(T_*, \theta) \leq MSE(T, \theta)$ for all $\theta \in \Theta$ and all candidate estimators T. We will not consider this problem here, and will only remark that in general this problem cannot be solved unless we restrict the class of estimators under consideration.

Instead, we will consider a slightly simpler version of the question posed, namely the following: for a given bias, can we make the variance of an estimator arbitrarily small? For example, if the bias is zero, and we have an unbiased estimator, is there a limit to how small the variance can be? The answer is given in the following theorem.

Theorem 3.9 (Cramér-Rao Lower Bound) *Let* $X_1, \ldots, X_n$ *be an iid sample from a regular parametric model* $f(\cdot; \theta)$, $\Theta \subseteq \mathbb{R}$. *Let* $T : \mathcal{X}^n \to \Theta$ *be an estimator of* θ, *for all* n. *Assume that:*

1. $\mathrm{Var}(T) < \infty$, *for all* $\theta \in \Theta$.
2. $\frac{\partial}{\partial \theta} \left[\int_{\mathcal{X}^n} f_{X_1,\ldots,X_n}(x_1, \ldots, x_n; \theta) dx \right] = \int_{\mathcal{X}^n} \frac{\partial}{\partial \theta} f_{X_1,\ldots,X_n}(x_1, \ldots, x_n; \theta) dx.$
3. $\frac{\partial}{\partial \theta} \left[\int_{\mathcal{X}^n} T(x_1, \ldots, x_n) f_{X_1,\ldots,X_n}(x_1, \ldots, x_n; \theta) dx_1 \ldots dx_n \right] =$

$$= \int_{\mathcal{X}^n} T(x_1, \ldots, x_n) \frac{\partial}{\partial \theta} f_{X_1,\ldots,X_n}(x_1, \ldots, x_n; \theta) dx_1 \ldots dx_n.$$

If we denote bias of T *by* $\beta(\theta) = \mathbb{E}(T) - \theta$, *then it holds that* $\beta(\theta)$ *is differentiable, and*

$$\mathrm{Var}(T) \geq \frac{\left(\beta'(\theta) + 1\right)^2}{n \int_{\mathcal{X}} \left(\frac{\partial}{\partial \theta} \log f(x; \theta)\right)^2 f(x; \theta) dx} = \frac{\left(\beta'(\theta) + 1\right)^2}{n \mathbb{E}\left[\frac{\partial}{\partial \theta} \log f(X_1; \theta)\right]^2}.$$

▶ **Remark 3.10** When X is a discrete random variable, the integrals above will be replaced by sums.

Even if the bias is equal to zero, the variance will still be bounded below by the inverse of the positive quantity $n \int_{\mathcal{X}} \left(\frac{\partial}{\partial \theta} \log f(x; \theta)\right)^2 f(x; \theta) dx =$ $n \mathbb{E}\left(\frac{\partial}{\partial \theta} \log f(X_1; \theta)\right)^2 = n I(\theta)$, and thus so will the MSE. For unbiased estimators, the variance (and hence the MSE) has the fundamental lower bound $1/n I(\theta)$. The quantity $I(\theta)$ is called the *Fisher information* or simply the *information*.[1] The presence of the term n^{-1} on the right-hand side of the Cramér–Rao inequality tells us that the best achievable variance when the sample size is n is of the order n^{-1}.

The good news is the following: if we are interested in looking only for *unbiased estimators*, and we find an unbiased estimator with variance $(n I(\theta))^{-1}$, then we know that we've found the best unbiased possible estimator in terms of MSE, regardless of the true value of θ.

---

[1] More generally, we may define

$$I_n(\theta) = \mathbb{E}\left[\frac{\partial}{\partial \theta} \log f_{X_1,\ldots,X_n}(X_1, \ldots, X_n; \theta)\right]^2$$

to be the Fisher information of a sample of size n. In the case of *iid* random variables, we have $I_n(\theta) = n I(\theta)$.

Proof of Theorem 3.9 First we prove the theorem in the special case $n = 1$. Define the random variable $U_1(\theta) = \frac{\partial}{\partial\theta}\log f(X_1;\theta)$. Since the probability model is regular, the support of f does not depend on θ. Therefore, by Assumption (3),

$$\mathbb{E}[U_1(\theta)] = \int_X \left(\frac{\partial}{\partial\theta}\log f(x;\theta)\right) f(x;\theta)dx = \int_X \frac{\frac{\partial}{\partial\theta}f(x;\theta)}{f(x;\theta)} f(x;\theta)dx$$

$$= \int_X \frac{\partial}{\partial\theta}f(x;\theta)dx = \frac{\partial}{\partial\theta}\int_X f(x;\theta)dx$$

$$= 0.$$

Therefore,

$$\mathrm{Var}[U_1(\theta)] = \mathbb{E}\left[U_1^2(\theta)\right] = \int_X \left(\frac{\partial}{\partial\theta}\log f(x;\theta)\right)^2 f(x;\theta)dx = I(\theta). \quad (3.1)$$

Again, since the support of f does not depend on θ and using Assumption (2),

$$\beta'(\theta) = \frac{\partial}{\partial\theta}\int_X T(x)f(x;\theta)dx - 1 = \int_X T(x)\frac{\partial}{\partial\theta}f(x;\theta)dx - 1$$

$$= \int_X T(x)\frac{\frac{\partial}{\partial\theta}f(x;\theta)}{f(x;\theta)} f(x;\theta)dx - 1$$

$$= \int_X T(x)\left(\frac{\partial}{\partial\theta}\log f(x;\theta)\right) f(x;\theta)dx - 1$$

$$= \mathbb{E}[T U_1(\theta)] - 1 = \left\{\mathbb{E}[T U_1(\theta)] - \mathbb{E}[T]\underbrace{\mathbb{E}[U_1(\theta)]}_{=0}\right\} - 1$$

$$= \mathrm{Cov}[U_1(\theta), T] - 1$$

$$\implies \mathrm{Cov}[U_1(\theta), T] = \beta'(\theta) + 1.$$

Now, using the correlation inequality[2] we have

$$\left|\frac{\mathrm{Cov}[U_1(\theta), T]}{\sqrt{\mathrm{Var}[U_1(\theta)]\mathrm{Var}[T]}}\right| \leq 1 \implies (\beta'(\theta) + 1)^2 \leq \mathrm{Var}[U_1(\theta)]\mathrm{Var}[T].$$

Finally, Eq. (3.1) allows us to conclude that

$$\mathrm{Var}[T] \geq \frac{(\beta'(\theta) + 1)^2}{\int_X \left(\frac{\partial}{\partial\theta}\log f(x;\theta)\right)^2 f(x;\theta)dx}.$$

---

[2] A consequence of the Cauchy–Schwarz inequality.

which proves the theorem when $n = 1$. For a more general n, define $U_i = \frac{\partial}{\partial\theta}\log f(X_i;\theta)$ and $U(\theta) = \sum_{i=1}^{n} U_i(\theta)$. Note that the $U_i(\theta)$ are independent and identically distributed as $U_1(\theta)$. Then, by linearity and independence, respectively,

$$\mathbb{E}[U(\theta)] = \sum_{i=1}^{n}\mathbb{E}[U_i(\theta)] = n\mathbb{E}[U_1(\theta)] = 0$$

$$\mathrm{Var}[U(\theta)] = \sum_{i=1}^{n}\mathrm{Var}[U_i(\theta)] = n\,\mathrm{Var}[U_1(\theta)] = n\int_{\mathcal{X}}\left(\frac{\partial}{\partial\theta}\log f(x;\theta)\right)^2 f(x;\theta)dx$$

$$\beta'(\theta) = \int_{\mathcal{X}^n} T(x_1,\ldots,x_n)\left(\sum_{i=1}^{n}\frac{\partial}{\partial\theta}\log f(x_i;\theta)\right)\prod_{i=1}^{n} f(x_i;\theta)dx_1\ldots dx_n - 1$$

$$= \mathrm{Cov}\,[U(\theta), T] - 1.$$

Applying the correlation inequality to $\mathrm{Cov}\,[U(\theta), T]$ then gives the result and completes the proof for general $n \geq 1$. □

Exercise 30 Let $X_1,\ldots,X_n \overset{iid}{\sim} Poisson(\lambda)$. Show that the estimator $\widehat{\lambda}_n = \overline{X}_n = \sum_{i=1}^{n} X_i/n$ of λ attains the Cramér–Rao lower bound.

▶ **Remark 3.11** Condition (3) of the theorem asks that we be able to interchange integration and differentiation. It can be checked on a case-by-case basis (for given f_x), or it can be replaced by any sufficient conditions on T and $f(x;\theta)$ allowing this to be the case. Here are two sets of conditions. Either would suffice for (3) to be true:

1. If T is such that we can write $f(x;\theta) = \exp\{\eta(\theta)T(x) - d(\theta) + S(x)\}$ with $\eta(\cdot)$ being differentiable on Θ with $\eta(\cdot)$ and $d(\cdot)$ functions on Θ as in exercise 23 page 52. In other words, if we have a one-parameter exponential family and the statistic T in question is its natural sufficient statistic (Bickel & Doksum [1], Proposition 3.4.1).
2. For $f(x;\theta)$ a density with $\theta \in \mathbb{R}$, and $T(x)$ being a real function, we have $\frac{\partial}{\partial\theta}\int_{\mathcal{X}} T(x)f(x;\theta)dx = \int_{\mathcal{X}} T(x)\frac{\partial}{\partial\theta}f(x;\theta)dx$ for all $\theta \in (a,b)$ if the following four conditions hold (Durrett [10], [Theorem 9.1]):
   (a) $\int_{\mathcal{X}} |T(x)|f(x;\theta)dx < \infty$ for all $\theta \in (a,b)$.
   (b) For any fixed $x \in \mathcal{X}$, $\frac{\partial}{\partial\theta}f(x;\theta)$ exists and is a continuous function of $\theta \in (a,b)$.
   (c) $\int_{\mathcal{X}} T(x)\frac{\partial}{\partial\theta}f(x;\theta)dx$ is continuous on (a,b).
   (d) $\int_{\mathcal{X}}\int_{a}^{b} \left|T(x)\frac{\partial}{\partial\theta}f(x;\theta)\right| d\theta dx < \infty$.

3.3   Methods for Constructing Estimators

Now we know a way to judge the quality of an estimator, and in some cases we also know what the best quality we can hope for is; but how can we propose candidate estimators? Any function from $\mathcal{X}^n \rightarrow \Theta$ is an estimator, so there is a bewildering variety of choice! We need general methods (or principles) that can be applied to any model in order to produce a candidate estimator. More ambitiously, we want methods that will generally yield reasonable estimators. If we have such methods, then we can study the properties of the estimators they induce.

3.3.1   The Method of Maximum Likelihood

Perhaps the most important method of point estimation is based on the notion of *likelihood*. We first give its rigorous definition, and then consider its intuitive interpretation.

Definition 3.12 (Likelihood for iid Collections)

Let $X_1, \ldots, X_n$ be a collection of independent and identically distributed random variables with density (or mass function) $f(x; \theta)$, where $\theta \in \mathbb{R}^p$. The likelihood of θ on the basis of $X_1, \ldots, X_n$ is defined as

$$L(\theta) = \prod_{i=1}^{n} f(X_i; \theta).$$

That is, the likelihood of θ is the joint density (or mass function) of the random variables $X_1, \ldots X_n$, evaluated at $(X_1, \ldots, X_n)$, but seen as a function of θ. Notice that the likelihood function is a *random function*, since it depends on the random sample $X_1, \ldots, X_n$. Strictly speaking, we should write $L_n(\theta)$ to denote the likelihood, in order to stress the fact that it depends on the sample size. Nevertheless, we will suppress the n index in general to simplify notation, with the exception of occasions where it is necessary for clarity.

The interpretation of the likelihood is easiest in the discrete case. In this case, the likelihood of θ is the probability of the observed sample $(X_1, \ldots, X_n)$, viewed as a function of θ. In other words, in the discrete case, the likelihood $L(\theta)$ is the answer to the question: *what is the probability of our observed sample when the parameter is taken to be equal to θ*[3]*?* When θ is unknown, it would seem that its most suitable estimate would be a value $\hat{\theta}$ that makes what we observed most probable—a value

---

[3] In the continuous case, a similar interpretation is feasible by considering a small neighbourhood around our sample: since $F(x + \epsilon/2; \theta) - F(x - \epsilon/2; \theta) \approx \epsilon f(x; \theta)$ as $\epsilon \downarrow 0$, we can think of $\epsilon^n L(\theta)$ as the approximate probability of a square neighbourhood of edge length ϵ centred around our sample, and viewed as a function of θ.

that is most compatible with our empirical observation. This motivates the definition of a maximum likelihood estimator.

Definition 3.13 (Maximum Likelihood Estimator)
Let $X_1, \ldots, X_n$ be an iid random sample from a distribution F_θ with density (or mass function) $f(x; \theta)$. Let $\hat{\theta}$ be such that

$$L(\theta) \leq L(\hat{\theta}), \qquad \forall\, \theta \in \Theta.$$

Then $\hat{\theta}$ is called a *maximum likelihood estimator* (MLE) of θ.

When there exists a unique maximum of the likelihood function, we speak of *the* maximum likelihood estimator $\hat{\theta} = \arg\max_{\theta \in \Theta} L(\theta)$. When the likelihood is a differentiable function of θ, we may determine the maximum likelihood estimator using differential calculus. A maximum of the function $L(\theta)$ must be a root of the equation

$$\nabla_\theta L(\theta) = 0$$

and so solving this equation will provide us with a candidate MLE. Before we declare a root $\hat{\theta}$ of this equation to actually be an MLE, we will first need to verify that this is indeed a maximum (and not a minimum! See Exercise 32, p. 74). If the likelihood is twice differentiable, this can be done by verifying that

$$-\nabla_\theta^2 L(\theta)\big|_{\theta=\hat{\theta}} \succ 0,$$

i.e. that minus the Hessian matrix is positive definite. When the parameter is one dimensional, this reduces to verifying that the second derivative is negative when evaluated at the root of the likelihood equation.

Notice that solving $\nabla_\theta L(\theta) = 0$ will involve the determination of the derivative of a product of n functions, which is a tedious calculation. To avoid this, we focus on maximising the *loglikelihood* $\ell(\theta) := \log L(\theta)$ instead of the likelihood itself. Since the log transform is monotone, the likelihood and the loglikelihood have precisely the same maxima and minima. But the advantage of the loglikelihood is that it is a sum rather than a product of n functions, making calculations straightforward:

$$\ell(\theta) = \log\left(\prod_{i=1}^n f(X_i; \theta)\right) = \sum_{i=1}^n \log f(X_i; \theta).$$

Again, if the loglikelihood function is twice differentiable, an MLE $\hat{\theta}$ of θ will satisfy

$$\nabla_\theta \ell(\theta)|_{\theta=\hat{\theta}} = 0 \quad \& \quad -\nabla_\theta^2 \ell(\theta)|_{\theta=\hat{\theta}} \succ 0.$$

Example 3.14 (Bernoulli MLE)

Let $X_1, \ldots, X_n \overset{iid}{\sim} Bern(p)$ and suppose we wish to use the maximum likelihood method to construct an estimator for $p \in (0, 1)$. The likelihood is

$$L(p) = \prod_{i=1}^n f(X_i; p) = \prod_{i=1}^n p^{X_i}(1-p)^{1-X_i} = p^{\sum_{i=1}^n X_i}(1-p)^{n-\sum_{i=1}^n X_i}.$$

Taking logarithms on both sides, we obtain the log likelihood function

$$\ell(p) = \log p \sum_{i=1}^n X_i + \log(1-p)\left(n - \sum_{i=1}^n X_i\right).$$

We notice that this function is indeed twice differentiable with respect to p, and calculate

$$\frac{d}{dp}\ell(p) = p^{-1}\sum_{i=1}^n X_i - (1-p)^{-1}\left(n - \sum_{i=1}^n X_i\right).$$

Solving for $\ell'(p) = 0$ with respect to p is equivalent to solving

$$p^{-1}\sum_{i=1}^n X_i - (1-p)^{-1}\left(n - \sum_{i=1}^n X_i\right) = 0$$

which can be seen to have the unique root $\frac{1}{n}\sum_{i=1}^n X_i = \overline{X}$. Call this $\hat{p}$. It is our candidate for an MLE, provided that it yields a maximum. Notice that

$$\frac{d^2}{dp^2}\ell(p) = -p^2\sum_{i=1}^n X_i - (1-p)^{-2}\left(n - \sum_{i=1}^n X_i\right)$$

which is always non-positive because $0 \leq \sum_{i=1}^n X_i \leq n$ almost surely and $p \in (0, 1)$. Hence $\hat{p} = \overline{X} = \frac{1}{n}\sum_{i=1}^n X_i$ is the unique MLE of p.   □

Example 3.15 (Exponential MLE)

Let $X_1, \ldots, X_n \overset{iid}{\sim} Exp(\lambda)$ and suppose we wish to use the maximum likelihood method to construct an estimator for $\lambda \in (0, \infty)$. The likelihood is

$$L(\lambda) = \prod_{i=1}^n f(X_i; \lambda) = \prod_{i=1}^n \lambda e^{-\lambda X_i} = \lambda^n \exp\left\{-\lambda \sum_{i=1}^n X_i\right\}.$$

Taking logarithms on both sides, we obtain the log likelihood function

$$\ell(\lambda) = n \log \lambda - \lambda \sum_{i=1}^{n} X_i.$$

We notice that this function is indeed twice differentiable with respect to p, and calculate

$$\frac{d}{d\lambda} \ell(\lambda) = n\lambda^{-1} - \sum_{i=1}^{n} X_i.$$

Solving for $\ell'(\lambda) = 0$ with respect to λ yields the unique root $\left(\frac{1}{n} \sum_{i=1}^{n} X_i\right)^{-1} = 1/\overline{X}$. Call this $\hat{\lambda}$. It is our candidate for an MLE, provided that it yields a maximum. Notice that

$$\frac{d^2}{d\lambda^2} \ell(\lambda) = -\frac{n}{\lambda^2}$$

which is always negative because $\lambda > 0$. Hence $\hat{\lambda} = \left(\frac{1}{n} \sum_{i=1}^{n} X_i\right)^{-1} = 1/\overline{X}$ is the unique MLE of λ.   □

Example 3.16 (Gaussian MLE)

Let $X_1, \ldots, X_n \overset{iid}{\sim} N(\mu, \sigma^2)$ and suppose we wish to use the maximum likelihood method to construct an estimator for $\theta = (\mu, \sigma^2) \in \mathbb{R} \times (0, \infty)$. The likelihood is

$$L(\mu, \sigma^2) = \prod_{i=1}^{n} f(X_i; \mu, \sigma^2) = \prod_{i=1}^{n} \frac{1}{\sqrt{2\pi\sigma^2}} \exp\left\{-\frac{(X_i - \mu)^2}{2\sigma^2}\right\}$$

$$= \left(\frac{1}{\sqrt{2\pi\sigma^2}}\right)^n \exp\left\{-\frac{\sum_{i=1}^{n} (X_i - \mu)^2}{2\sigma^2}\right\}.$$

Taking logarithms on both sides, we obtain the log likelihood function

$$\ell(\mu, \sigma^2) = -\frac{n}{2} \log(2\pi\sigma^2) - \frac{1}{2\sigma^2} \sum_{i=1}^{n} (X_i - \mu)^2.$$

We notice that all second order derivatives with respect to μ and σ^2 exist, and calculate

$$\frac{\partial}{\partial \mu} \ell(\mu, \sigma^2) = \frac{1}{\sigma^2} \sum_{i=1}^{n} (X_i - \mu)$$

$$\frac{\partial}{\partial \sigma^2} \ell(\mu, \sigma^2) = -\frac{n}{2\sigma^2} + \frac{1}{2\sigma^4} \sum_{i=1}^{n} (X_i - \mu)^2.$$

Solving for $\nabla_{(\mu,\sigma^2)} \ell(\mu, \sigma^2) = 0$ with respect to (μ, σ^2) yields a system of two equations in two unknowns. The unique root of this system can be seen to be $\left(\overline{X}, n^{-1} \sum_{i=1}^{n} (X_i - \overline{X})^2\right)$. Call this

$(\hat{\mu}, \hat{\sigma}^2)$. It is our candidate for an MLE, provided that it yields a maximum. Notice that

$$\frac{\partial^2}{\partial \mu^2} \ell(\mu, \sigma^2) = -\frac{n}{\sigma^2}, \quad \frac{\partial^2}{\partial (\sigma^2)^2} \ell(\mu, \sigma^2) = \frac{n}{2\sigma^4} - \frac{1}{\sigma^6} \sum_{i=1}^{n} (X_i - \mu)^2$$

$$\frac{\partial^2}{\partial \mu \partial \sigma^2} \ell(\mu, \sigma^2) = \frac{\partial^2}{\partial \sigma^2 \partial \mu} \ell(\mu, \sigma^2) = -\frac{\sum_{i=1}^{n} (X_i - \mu)}{\sigma^4} = \frac{n\mu - n\overline{X}}{\sigma^4}.$$

Evaluating these second derivatives at $(\hat{\mu}, \hat{\sigma}^2)$ yields

$$\frac{\partial^2}{\partial \mu^2} \ell(\mu, \sigma^2) \bigg|_{(\mu, \sigma^2) = (\hat{\mu}, \hat{\sigma}^2)} = -\frac{n}{\hat{\sigma}^2}, \quad \frac{\partial^2}{\partial (\sigma^2)^2} \ell(\mu, \sigma^2) \bigg|_{(\mu, \sigma^2) = (\hat{\mu}, \hat{\sigma}^2)} = \frac{n}{2\hat{\sigma}^4} - \frac{n\hat{\sigma}^2}{\hat{\sigma}^6} = -\frac{n}{2\hat{\sigma}^4}$$

$$\frac{\partial^2}{\partial \mu \partial \sigma^2} \ell(\mu, \sigma^2) \bigg|_{(\mu, \sigma^2) = (\hat{\mu}, \hat{\sigma}^2)} = \frac{\partial^2}{\partial \sigma^2 \partial \mu} \ell(\mu, \sigma^2) \bigg|_{(\mu, \sigma^2) = (\hat{\mu}, \hat{\sigma}^2)} = \frac{n\hat{\mu} - n\hat{\mu}}{\hat{\sigma}^4} = 0.$$

We conclude that the matrix $\left[-\nabla^2_{(\mu, \sigma^2)} \ell(\mu, \sigma^2) \bigg|_{(\mu, \sigma^2) = (\hat{\mu}, \hat{\sigma}^2)} \right]$ is diagonal. To show that it is positive definite, it suffices to show that its two diagonal elements are positive, which is true since $\hat{\sigma}^2$ is positive with probability one. Therefore, the unique MLE of (μ, σ^2) is given by

$$(\hat{\mu}, \hat{\sigma}^2) = \left(\overline{X}, \frac{1}{n} \sum_{i=1}^{n} (X_i - \overline{X})^2 \right).$$

□

There are situations where we might not be interested in estimating θ itself, but rather some transformation $\phi = g(\theta)$. If g is a bijection, we do not need to repeat the entire estimation process, since the maxima of a function are equivariant to a reparametrisation of its domain.

Proposition 3.17 (Bijective Equivariance of the MLE) *Let $\{ f(\cdot; \theta) : \theta \in \Theta \}$ be a parametric model, where $\Theta \subseteq \mathbb{R}^p$. Suppose that $\hat{\theta}$ be an MLE of θ, on the basis of a random sample $X_1, \ldots, X_n$ from $f(x; \theta)$. Let $g : \Theta \to \Phi \subseteq \mathbb{R}^p$ be a bijection. Then, $\hat{\phi} = g(\hat{\theta})$ is an MLE of $\phi = g(\theta)$.*

Proof Define $h(x; \phi) = f(x; g^{-1}(\phi))$, and note that h is a well-defined function, because $g^{-1} : \Phi \to \Theta$ is well defined. The function $h(x; \phi)$ is simply the density/frequency of X_i under the parametrisation given by $\phi \in \Phi$. An MLE of ϕ, say $\hat{\phi}$ must satisfy

$$\prod_{i=1}^{n} h(X_i; \phi) \le \prod_{i=1}^{n} h(X_i; \hat{\phi}), \quad \forall \phi \in \Phi.$$

Let $\hat{\theta}$ be an MLE of θ, and let $\hat{\phi} = g(\hat{\theta})$. Let $\phi \in \Phi$ be arbitrary and observe that

$$\prod_{i=1}^{n} h(X_i; \phi) = \prod_{i=1}^{n} f(X_i; g^{-1}(\phi)) \leq \prod_{i=1}^{n} f(X_i; \hat{\theta})$$

$$= \prod_{i=1}^{n} f(X_i; g^{-1}(\hat{\phi}))$$

$$= \prod_{i=1}^{n} h(X_i; \hat{\phi}),$$

which proves the proposition.   □

Example 3.18

Let $X_1, .., X_n \overset{iid}{\sim} \mathcal{N}(\mu, 1)$, and suppose we are interested in estimating the probability $\mathbb{P}[X_1 \leq x]$, for a given $x \in \mathbb{R}$. We note that

$$\mathbb{P}[X_1 \leq x] = \mathbb{P}[X_1 - \mu \leq x - \mu] = \Phi(x - \mu),$$

where Φ is the standard normal CDF (see Lemma 1.32, p. 22). But the mapping $\mu \mapsto \Phi(x - \mu)$ is a bijection because Φ is monotone; thus, the MLE of $\mathbb{P}[X_1 \leq x]$ is $\Phi(x - \hat{\mu})$, where $\hat{\mu}$ is the MLE of μ (which from our previous example is $\hat{\mu} = \overline{X}$).   □

Example 3.19 (Usual vs Natural Parameter in Exponential Families)

Let $X_1, \ldots, X_n \overset{iid}{\sim} f$, with

$$f(x) = \exp \{\phi T(x) - \gamma(\phi) + S(x)\}, \qquad x \in \mathcal{X}$$

where $\phi \in \Phi \subseteq \mathbb{R}$ is the natural parameter. Now suppose that we can also write $\phi = \eta(\theta)$ for $\theta \in \Theta$ is the usual parameter, and $\eta : \Theta \to \Phi$ some differentiable 1-1 mapping (and so $\gamma(\phi) = \gamma(\eta(\theta)) = d(\theta)$, for $d = \gamma \circ \eta$). In this form, the exponential family density/frequency will take the form:

$$\exp \{\phi T(x) - \gamma(\phi) + S(x)\} = \exp \{\eta(\theta)T(x) - d(\theta) + S(x)\}.$$

Now, Proposition 3.17 (p. 72) implies that if $\hat{\theta}$ is the MLE of θ, then $\eta(\hat{\theta})$ is the MLE of $\phi = \eta(\theta)$. The converse is also true: if $\hat{\phi}$ is the unique MLE of ϕ, then $\eta^{-1}(\hat{\phi})$ is the unique MLE of $\theta = \eta^{-1}(\phi)$. For concrete examples, see Examples 1.24 (p. 18) and 1.26 (p. 18).   □

Exercise 31 Let $X_1, \ldots, X_n \overset{iid}{\sim} Exp(\lambda)$, where $n > 2$, and let $\hat{\lambda}_n$ be the MLE of λ on the basis of the sample.

1. Show that $\mathbb{E}_\lambda(\hat{\lambda}_n) = \lambda n/(n-1)$, and find a new estimator $\hat{\lambda}_n^U$ that is unbiased for λ. Hint: use the fact that $Z = \sum_{i=1}^n X_i \sim Gamma(n, \lambda)$.
2. Show that $Var_\lambda(\hat{\lambda}_n) = n^2 \lambda^2/\left((n-1)^2 (n-2)\right)$.
3. Does the estimator $\hat{\lambda}_n^U$ attain the Cramér–Rao bound?
4. Determine the MLE $\hat{\theta}_n$ and the Cramér–Rao bound associated with the parameter $\theta = 1/\lambda$. Can we use Proposition 3.17?

    Compare the variance of $\hat{\theta}_n$ and the obtained Cramér–Rao bound.

There are situations where differential calculus will not be applicable, and other approaches will be necessary. This can happen, for example, in models with discrete parameter spaces Θ or in models where the support of $L(\theta)$ depends on θ. If θ is one-dimensional, one can sometimes employ direct inspection in order to determine the MLE.

Example 3.20

Let $X_1, \ldots, X_n \overset{iid}{\sim} Unif(0, \theta)$. The likelihood is

$$L(\theta) = \theta^{-n} \prod_{i=1}^n \mathbf{1}\{0 \le X_i \le \theta\} = \theta^{-n}\mathbf{1}\{\theta \ge X_{(n)}\}\mathbf{1}\{X_{(1)} > 0\}.$$

Hence if $\theta < X_{(n)}$ the likelihood is zero. In the domain $[X_{(n)}, \infty)$, the likelihood is a decreasing function of θ. Hence $\hat{\theta} = X_{(n)}$.   □

Exercise 32 (Minimum Likelihood) Let X be a discrete random variable taking the values

$$\begin{cases} 0 & \text{with probability } 6\theta^2 - 4\theta + 1; \\ 1 & \text{with probability } \theta - 2\theta^2; \\ 2 & \text{with probability } 3\theta - 4\theta^2, \end{cases}$$

where $\theta \in [0, 1/2]$. Determine the maximum likelihood estimator on the basis of a sample of size 1, X_1. What do you observe?

Exercise 33 (Conditional Likelihood) Let $X_1, \ldots, X_m \overset{iid}{\sim} Exp(\lambda)$, where $\lambda > 0$. How does the MLE of λ change if we somehow are told that all X_i overshot their mean? (in mathematical terms, conditional on the event $\{X_i > \mathbb{E}[X_i], i = 1 \ldots, n\}$). Note that as with example 3.20, the support of the conditional distribution depends on the true parameter value λ.

3.3.2 Maximum Likelihood in Exponential Families

With the exception of the uniform distribution, all the examples of probability models we have seen so far on the use of the maximum likelihood method are special cases of exponential families. It is then natural to wonder whether some general results can be obtained on the use of the method of maximum likelihood in an arbitrary parametric model that is a member of the exponential family.

It was no accident that the MLE existed and was unique in Examples 3.14 (p. 70), 3.15 (p. 70), and 3.16 (p. 71). This is a general phenomenon for models that are exponential families. We will consider here the one-parameter case for simplicity.

Proposition 3.21 (One-Parameter Exponential Family MLE) *Let $X_1, \ldots, X_n$ be an iid sample from a distribution with density/frequency in a one-parameter exponential family,*

$$f(x; \phi) = \exp\{\phi T(x) - \gamma(\phi) + S(x)\}, \qquad x \in \mathcal{X}, \phi \in \Phi$$

with a parameter space $\Phi \subset \mathbb{R}$ that is an open set and T a non-constant function. If the MLE $\hat{\phi}$ of ϕ exists, then it is unique, and is given by the unique solution to the equation

$$\gamma'(u) = \overline{T},$$

with respect to u. Here, $\overline{T} = \frac{1}{n} \sum_{i=1}^{n} T(X_i)$.

Proof The likelihood of ϕ on the basis of the sample $X_1, \ldots, X_n$ is

$$L(\phi) = \prod_{i=1}^{n} e^{\phi T(X_i) - \gamma(\phi) + S(X_i)}$$

from which we deduce that the loglikelihood is

$$\ell(\phi) = \log L(\phi) = -n\gamma(\phi) + \sum_{i=1}^{n} S(X_i) + \phi \sum_{i=1}^{n} T(X_i) = -n\gamma(\phi) + \sum_{i=1}^{n} S(X_i) + n\phi \overline{T}.$$

Since $\gamma(\cdot)$ is twice differentiable, we may also differentiate ℓ twice. Doing so, we obtain that

$$\ell''(\phi) = -n\gamma''(\phi) = -\text{Var}\left[\sum_{i=1}^{n} T(X_i)\right] \leq 0,$$

where the last equality comes from Proposition 2.11 (p. 51). Since the second derivative is negative for all ϕ, the function $\ell(\phi)$ is concave. Thus if it attains a maximum in Φ this must be the unique maximum, which proves that the MLE is unique. Since Φ is open, this maximum $\hat{\phi}$ of $\ell(\phi)$ must uniquely solve the equation $\ell'(\phi) = 0$ with respect to ϕ, or equivalently it must uniquely satisfy

$$\gamma'(\hat{\phi}) = \overline{T}.$$

$\square$

▶ **Remark 3.22 (Usual Parametrisation)**   If $\phi = \eta(\theta)$ for a bijection η, then the MLE of θ is also unique, if it exists, by Proposition 3.17 (p. 72).

Exercise 34 (Cramér–Rao Bound and Exponential Families) Let $f(x; \theta) = \exp(\eta(\theta)T(x) - d(\theta) + S(x))$ be a non-degenerate exponential family such that
- The parameter space $\Theta \subseteq \mathbb{R}$ is open;
- $T(X)$ is not a constant function (i.e. $\text{Var}_\theta[T(X)] > 0$) for all θ;
- The function $\eta : \Theta \to \mathbb{R}$ is a twice differentiable injection with non-vanishing first derivative.

Let $X_1, \ldots, X_n \overset{iid}{\sim} f(x; \theta)$. Suppose that the MLE $\widehat{\theta}_n$ of θ exists and has finite variance for all $\theta \in \Theta$. Prove that $\widehat{\theta}_n$ attains the Cramér–Rao bound (for all $\theta \in \Theta$) if and only if $h(\theta) = d'(\theta)/\eta'(\theta)$ is an affine function ($h(\theta) = \alpha\theta + \beta$ for some $\alpha, \beta \in \mathbb{R}$).

Hint: at some point in the proof of the Cramér–Rao theorem, we use a certain inequality. The Exercise 23 (p. 52) will be useful to show that $\widehat{\theta}_n$ corresponds to a maximum.

3.3.3   Large Sample Properties of Maximum Likelihood

Going back to Example 3.16 (p. 71), we recall that the maximum likelihood estimator for the parameter (μ, σ^2) of a Gaussian distribution, based on an iid sample $X_1, \ldots, X_n$, is

$$(\hat{\mu}_n, \hat{\sigma}_n^2) = \left(\bar{X}, \frac{1}{n}\sum_{i=1}^{n}(X_i - \bar{X})^2\right) = \left(\bar{X}, \frac{n-1}{n}S_n^2\right),$$

where $S_n^2 = \frac{1}{n-1}\sum_{i=1}^{n}(X_i - \bar{X})^2$. Using Proposition 2.7 (p. 45) and Corollary 2.8 (p. 48) we thus have a complete description of the probabilistic behaviour of these estimators:
- The MLE of μ, $\hat{\mu}_n$, is unbiased for all n. Its distribution is normal for all n, with variance σ^2/n. Therefore, the mean squared error is, in fact, exactly σ^2/n, regardless of the vale of μ.

- The MLE of σ^2, $\hat{\sigma}_n^2$ is biased for all n. By Corollary 2.8 (p. 48) its bias is equal to:

$$bias(\hat{\sigma}_n^2, \sigma^2) = \mathbb{E}[\hat{\sigma}_n^2] - \sigma^2 = \mathbb{E}\left[\frac{n-1}{n}S^2\right] - \sigma^2 = \frac{n-1}{n}\sigma^2 - \sigma^2 = -\frac{1}{n}\sigma^2.$$

Therefore, $\hat{\sigma}_n^2$ underestimates σ^2, though asymptotically the bias reduces to zero. The distribution of $\hat{\sigma}_n^2$ is the same as that of a chi-square random variable, multiplied by σ^2/n. That is:

$$\frac{n}{\sigma^2}\hat{\sigma}_n^2 \sim \chi_{n-1}^2.$$

Consequently, the mean squared error of $\hat{\sigma}_n^2$ is

$$MSE(\hat{\sigma}_n^2, \sigma^2) = bias^2(\hat{\sigma}_n^2, \sigma^2) + Var[\hat{\sigma}_n^2] = \left(-\frac{\sigma^2}{n}\right)^2 + \frac{2(n-1)\sigma^4}{n^2} = \frac{(2n-1)\sigma^4}{n^2}.$$

Exercise 35 Let $X_1, \ldots, X_n \overset{iid}{\sim} N(\mu, \sigma^2)$ where both parameters are unknown ($n > 1$). We can estimate σ^2 by $S_n^2 = \frac{1}{n-1}\sum_{i=1}^{n}(X_i - \overline{X})^2$, or by the MLE given by $\hat{\sigma}_n^2 = (n-1)S_n^2/n$.
1. Which of the two estimators is preferable in terms of mean squared error?
2. More generally, consider estimators of the form aS_n^2, where $a \in \mathbb{R}$. Which is the optimal choice of a in terms of mean squared error?

We can gain a visual understanding of the behaviour of the MLE in the Gaussian case by looking at Figs. 3.1 (p. 78) and 3.2 (p. 79). These illustrate the sampling fluctuations of the MLE around the true parameter value, and how these change as the sample size increases. We note that, as n increases, the realisations of the MLE concentrate more and more around the true parameter values. This is no accident: the mean squared error both for $\hat{\mu}$ and for $\hat{\sigma}^2$ is decreasing in n, with a limit of 0 as $n \to \infty$. Therefore, both estimators are consistent (recall Lemma 3.6, p. 63).

The normal case is special in that we can determine the exact sampling distribution of the maximum likelihood estimator and determine the mean squared error for every n. This gives us all the information we need in terms of the performance of the estimator.

Unfortunately, we are not always as lucky with models other than the normal distribution. The exact sampling distribution of the MLE is often not available, nor is the exact value of the MSE. As we saw in Sect. 2.4, when we cannot determine a sampling distribution exactly, we need to resort to approximations using the notion of convergence in distribution. In fact, we saw that for one-parameter exponential families the approximate distribution of the natural sufficient statistic $\overline{T}_n$ is normal (Corollary 2.24, p. 56). Since the MLE in a one-parameter exponential family is given by the solution of an equation involving $\overline{T}_n$ (see Proposition 3.21, p. 75), one might conjecture that perhaps the asymptotic distribution of the MLE in a one-

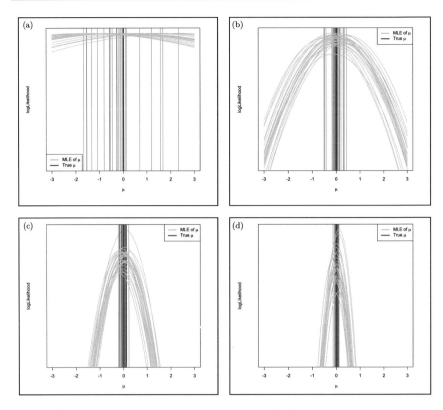

Fig. 3.1 Illustration of the random fluctuations of the loglikelihood function and its maximum (the MLE). We consider the estimation of the mean μ of a normal distribution with a known variance equal to 1. We generate 25 iid samples of size n, say $\{X_{i,1}, \ldots, X_{i,n}\}_{i=1}^{25}$, from an $N(\mu, 1)$ where $\mu = 0$, and each time plot the loglikelihood function $\ell_i(\mu) = \ell(\mu; X_{i,1}, \ldots, X_{i,n})$, where $i = 1, 2, \ldots, 25$, and the corresponding MLE. We do this for four sample sizes: $n = 1$, $n = 20$, $n = 100$, $n = 400$. We observe how the likelihood functions become gradually more curved as n increases, and so their maximum fluctuates less and less from replication to replication. We also notice that the maxima tend to concentrate around the true value of μ as n increases. The y-axis values have been removed since they are unimportant in an absolute sense in the determination of the MLE. (**a**) Loglikelihood functions for the mean parameter corresponding to 25 replications of an iid $N(0, 1)$ sample of size 1. (**b**) Loglikelihood functions for the mean parameter corresponding to 25 replications of an iid $N(0, 1)$ sample of size 20. (**c**) Loglikelihood functions for the mean parameter corresponding to 25 replications of an iid $N(0, 1)$ sample of size 100. (**d**) Loglikelihood functions for the mean parameter corresponding to 25 replications of an iid $N(0, 1)$ sample of size 450

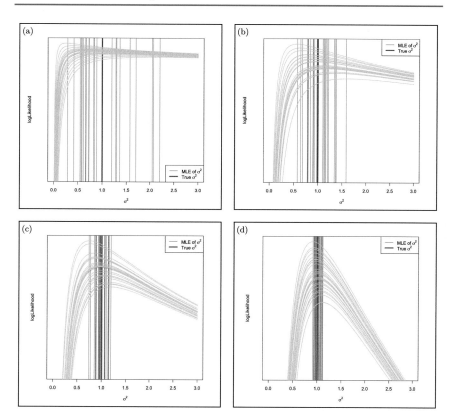

Fig. 3.2 Illustration of the random fluctuations of the loglikelihood function and its maximum (the MLE). We consider the estimation of the variance σ^2 of a normal distribution with a known mean equal to 0. We generate 25 iid samples of size n, say $\{X_{i,1}, \ldots, X_{i,n}\}_{i=1}^{25}$, from an $N(0, \sigma^2)$ where $\sigma^2 = 1$, and each time plot the log likelihood function $\ell_i(\sigma^2) = \ell(\sigma^2; X_{i,1}, \ldots, X_{i,n})$, where $i = 1, 2, \ldots, 25$, and the corresponding MLE. We do this for four sample sizes: $n = 10$, $n = 50$, $n = 150$, $n = 450$. We observe how the likelihood functions become gradually more curved as n increases, and so their maximum fluctuates less and less from replication to replication. We also notice that the maxima tend to concentrate around the true value of σ^2 as n increases—in fact, as n increases, it looks as though the distribution of the maxima is gradually becoming symmetric around σ^2. The y-axis values have been removed since they are unimportant in an absolute sense in the determination of the MLE. (**a**) Loglikelihood functions for the variance parameter corresponding to 25 replications of an iid $N(0, 1)$ sample of size 10. (**b**) Loglikelihood functions for the variance parameter corresponding to 25 replications of an iid $N(0, 1)$ sample of size 50. (**c**) Loglikelihood functions for the variance parameter corresponding to 25 replications of an iid $N(0, 1)$ sample of size 150. (**d**) Loglikelihood functions for the variance parameter corresponding to 25 replications of an iid $N(0, 1)$ sample of size 450

parameter exponential family is also normal (because, if the solution of the equation depends smoothly on $\overline{T}_n$, then the delta method (Theorem 2.27, p. 57) could be invoked). This is indeed the case.

Theorem 3.23 *Let $X_1, \ldots, X_n$ be an iid sample from a distribution with density (or mass function) $f(x; \phi_0)$ which belongs to a non-degenerate one-parameter exponential family,*

$$f(x; \phi) = \exp\{\phi T(x) - \gamma(\phi) + S(x)\}, \qquad x \in \mathcal{X}, \ \phi \in \Phi,$$

such that T is not a constant function. Assume that the parameter space $\Phi \subset \mathbb{R}$ is an open set (recall that, among others, this implies that the function $\gamma(\cdot)$ is twice differentiable). Let $\hat{\phi}_n$ be the maximum likelihood estimator of ϕ_0, assumed to exist. Then,

$$0 < \frac{1}{\gamma''(\phi_0)} < \infty$$

and

$$\sqrt{n}(\hat{\phi}_n - \phi_0) \xrightarrow{d} N\left(0, \frac{1}{\gamma''(\phi_0)}\right).$$

▶ **Remark 3.24 (Non-Degeneracy)** To say that a distribution is *non-degenerate* (as the theorem requires) means that it does not assign probability 1 to a single value $x \in \mathcal{X}$.

Proof of Theorem 3.23 Under the conditions of the Theorem, Proposition 3.21 (p. 75) implies that the MLE $\hat{\phi}_n$ is unique for all n. Furthermore, Proposition 2.11 (p. 51) implies that

$$\gamma''(\phi) = \frac{1}{n} \text{Var}\left[\sum_{i=1}^{n} T(X_i)\right] \in [0, \infty)$$

which proves that $0 \leq \frac{1}{\gamma''(\phi_0)} < \infty$ for all $\phi \in \Phi$. To prove that $\gamma''(\phi) > 0$ (strict inequality) we remark that it must be that $\text{Var}(T_i) > 0$. Because if $\text{Var}(T_i) = 0$, then $\mathbb{P}[T_i = \mathbb{E}(T_i)] = 1$ (by Chebyshev's inequality, Lemma A.4, p. 159) which means that X_i is almost surely constant or $T(\cdot)$ is a constant function on $\mathcal{X}$. Either of these contradicts our assumptions (that the exponential family in question is non-degenerate and that T is non-constant). We thus conclude that $0 < \frac{1}{\gamma''(\phi)} < \infty$ for all $\phi \in \Phi$.

Now, since Φ is open, the unique maximum $\hat{\phi}_n$ of $\ell(\phi)$ must uniquely solve the equation $\ell'(\phi) = 0$ with respect to ϕ, or equivalently it must uniquely satisfy

$$\gamma'(\hat{\phi}_n) = \overline{T}.$$

Since γ' is continuously differentiable (by assumption (1)) and we've shown that $\gamma'' > 0$, the inverse function theorem[4] implies that there exists an open ball of radius $\epsilon > 0$ centred at $\gamma'(\phi_0)$, say $B_\epsilon(\gamma'(\phi_0)) = \{y \in \mathbb{R} : |y - \gamma'(\phi_0)| < \epsilon\}$, such that $g(\cdot) = [\gamma']^{-1}(\cdot)$ exists on $B_\epsilon(\gamma'(\phi_0))$ and is itself differentiable with a continuous first derivative, which is in fact given explicitly by

$$g'(y) = \{[\gamma']^{-1}\}'(y) = \frac{1}{\gamma''([\gamma']^{-1}(y))} = \frac{1}{\gamma''(g(y))}.$$

By convention, we may define g to be zero outside of $B_\epsilon(\gamma'(\phi_0))$.
Now, Corollary 2.24 (p. 56) implies that[5]

$$\sqrt{n}(\overline{T} - \gamma'(\phi_0)) \xrightarrow{d} N(0, \gamma''(\phi_0)).$$

If we define $\tilde{\phi}_n = g(\overline{T})$, then delta method (Theorem 2.27, p. 57) implies

$$\sqrt{n}(\tilde{\phi}_n - \phi_0) = \sqrt{n}(g(\overline{T}) - g(\gamma'(\phi_0))) \xrightarrow{d} N\left(0, \gamma''(\phi_0) \times [g'(\gamma'(\phi_0))]^2\right).$$

But, by the inverse function theorem, since $g(y) = [\gamma']^{-1}(y)$,

$$g'(y) = \frac{1}{\gamma''(g(y))} \implies g'(\gamma'(\phi_0)) = \frac{1}{\gamma''(g(\gamma'(\phi_0)))} = \frac{1}{\gamma''(\phi_0)}$$

and so we conclude that

$$\sqrt{n}(\tilde{\phi}_n - \phi_0) \xrightarrow{d} N\left(0, \frac{1}{\gamma''(\phi_0)}\right).$$

To complete the proof, suppose that we can show that

$$\sqrt{n}(\hat{\phi}_n - \tilde{\phi}_n) \xrightarrow{p} 0.$$

---

[4]Recall the inverse function theorem: let $h(x) : \mathbb{R} \to \mathbb{R}$ be continuously differentiable, with a non-zero derivative at a point $x_o \in \mathbb{R}$. Then, there exists an $\varepsilon > 0$ such h^{-1} exists and is continuously differentiable on $(h(x_0) - \epsilon, h(x_0) + \epsilon)$, and in fact $(h^{-1})'(y) = [h'(h^{-1}(y))]^{-1}$ for $|y - h(x_0)| < \varepsilon$.

[5]Remember: since $\overline{T}$ is the sum of the iid terms $T_1, \ldots, T_n$, each satisfying $\mathrm{Var}(T(X_i)) = \gamma''(\phi_0) < \infty$ and $\mathbb{E}[T(X_i)] = \gamma'(\phi_0)$, so the central limit theorem implies $\sqrt{n}(\overline{T} - \gamma'(\phi_0)) \xrightarrow{d} N(0, \gamma''(\phi_0))$.

Then, Slutsky's theorem (Theorem 2.26, p. 57) will imply that $\sqrt{n}(\hat{\phi}_n - \phi_0) \xrightarrow{d}$ $N\left(0, \frac{1}{\gamma''(\phi_0)}\right)$ and prove the theorem.[6] Note, however, that

$$\overline{T} \in B_\epsilon(\gamma'(\phi_0)) \implies \hat{\phi}_n = \tilde{\phi}_n \implies \sqrt{n}(\tilde{\phi}_n - \hat{\phi}_n) = 0,$$

because $\hat{\phi}_n = g(\overline{T}) = \tilde{\phi}_n$ when $\overline{T} \in B_\epsilon(\gamma'(\phi_0))$. Therefore, if $\delta > 0$,

$$\sqrt{n}|\tilde{\phi}_n - \hat{\phi}_n| > \delta \implies \overline{T} \notin B_\epsilon(\gamma'(\phi_0)) \implies |\overline{T} - \gamma'(\phi_0)| > \epsilon,$$

and consequently,

$$\mathbb{P}[\sqrt{n}|\tilde{\phi}_n - \hat{\phi}_n| > \delta] \leq \mathbb{P}[|\overline{T} - \gamma'(\phi_0)| > \epsilon] \xrightarrow{n \to \infty} 0.$$

The convergence to zero follows from the weak law of large numbers.[7] This proves that $\sqrt{n}(\hat{\phi}_n - \tilde{\phi}_n) \xrightarrow{p} 0$ and completes the proof. □

▶ **Remark 3.25 (Asymptotic Variance and the Cramér–Rao Bound)** The theorem can be interpreted as saying that, for large n, the MLE $\hat{\phi}$ is approximately $N(\phi_0, [n\gamma''(\phi_0)]^{-1})$. We notice that the asymptotic mean of the MLE is equal to the true parameter, so that the asymptotic bias is zero. Furthermore, we note that

$$\mathbb{E}[(\ell'(\phi))^2] = \mathbb{E}\left\{\left[\frac{\partial}{\partial \phi}\left(\phi\tau(X_1, \ldots, X_n) - n\gamma(\phi)\right)\right]^2\right\}$$

$$= \mathbb{E}\left[\left(\tau(X_1, \ldots, X_n) - n\gamma'(\phi)\right)^2\right]$$

$$= \mathrm{Var}[\tau(X_1, \ldots, X_n)]$$

$$= n\gamma''(\phi).$$

Now recall the Cramér–Rao lower bound (Theorem 3.9, p. 65) on the variance of an estimator. It stated that no unbiased estimator can have variance lower than the inverse of the left-hand side of the equation above. But we have just proved that the inverse of the right-hand side is the asymptotic variance of the MLE. It follows that, at least for large sample size n, the maximum likelihood estimator of ϕ attains a performance which is close to optimal. This explains why the method of maximum likelihood is so central to point estimation.

---

[6]To see this, use Slutsky's theorem with $X_n = \sqrt{n}(\tilde{\phi}_n - \phi_0)$, $Y_n = \sqrt{n}(\hat{\phi}_n - \tilde{\phi}_n)$ and the continuous mapping being $(X_n, Y_n) \mapsto (X_n + Y_n)$.

[7]Since $\overline{T}$ is the mean of the iid terms $T(X_1), \ldots, T(X_n)$, each satisfying $\mathrm{Var}(T(X_i)) = \gamma''(\phi_0) < \infty$ and $\mathbb{E}[T(X_i)] = \gamma'(\phi_0)$, the Law of Large Numbers implies that $\overline{T} \xrightarrow{p} \gamma'(\phi_0)$

Corollary 3.26 (Consistency of the MLE in Exponential Families) *In the same setup and with the same conditions as in Theorem 3.23 (p. 80), we have*

$$\hat{\phi}_n \xrightarrow{p} \phi_0, \qquad as\ n \to \infty.$$

Proof Define $Y_n = n^{-1/2}$, $X_n = \sqrt{n}(\hat{\phi}_n - \phi_0)$ and $g : \mathbb{R} \times \mathbb{R} \to \mathbb{R}$ as $g(x, y) = xy$. Then, Theorem 3.23 (p. 80) combined with Slutsky's theorem (Theorem 2.26, p. 57) imply that

$$g(X_n, Y_n) = (\hat{\phi}_n - \phi_0) \xrightarrow{d} 0.$$

Consequently, Lemma 2.20 (p. 55) implies that $(\hat{\phi}_n - \phi_0) \xrightarrow{p} 0$ and the proof is complete. □

Notice that $\gamma''(\phi) = -\ell''(\phi)$ is minus the second derivative of the loglikelihood. Even though the loglikelihood is a random function, in the case of an exponential family its second derivative is a deterministic function of ϕ. What is the interpretation of this function? Recall that the second derivative of a function at some point ϕ_0 describes the curvature of the function at that point. Therefore, Theorem 3.23 (p. 80) tells us that the asymptotic variance of the MLE is related to the curvature of the loglikelihood at the true parameter ϕ_0. A moment of thought should reveal that this is quite intuitive: the more flat the loglikelihood is around the true parameter, the more "uncertain" its maximum will be: a small perturbation in the loglikelihood (e.g. due to sample variation) will yield a large perturbation of its maximum, due to flatness (hence high variance). On the other hand, if the loglikelihood is very pointy, we expect that the maximum will not change very much when the loglikelihood is perturbed (low variance). This phenomenon is clearly visible in Figs. 3.1 (p. 78) and 3.2 (p. 79), where we can see that the dispersion of the MLEs reduces as the curvature of the loglikelihood increases.

The asymptotic distribution for the usual parameter $\theta = \eta^{-1}(\phi)$ of an exponential family will now follow as a corollary.

Corollary 3.27 *Let $X_1, \ldots, X_n$ be an iid sample from a distribution with density (or mass function) $f(x; \theta_0)$ which belongs to a non-degenerate one-parameter exponential family,*

$$f(x; \theta) = \exp\{\eta(\theta)T(x) - d(\theta) + S(x)\}, \qquad x \in \mathcal{X}, \theta \in \Theta$$

Assume that:
1. *The parameter space $\Theta \subset \mathbb{R}$ is an open set.*
2. *The function $\eta(\cdot)$ is a twice differentiable bijection between Θ and $\Phi = \eta(\Theta)$ with non-vanishing derivative.*

3. The function $T(x) : \mathcal{X} \to \mathbb{R}$ is not a constant.
Let $\hat{\theta}_n$ be the maximum likelihood estimator of θ_0, assumed to exist. Then,

$$\sqrt{n}(\hat{\theta}_n - \theta_0) \xrightarrow{d} N\left(0, \frac{[\eta'(\theta_0)]}{d''(\theta_0)\eta'(\theta_0) - d'(\theta_0)\eta''(\theta_0)}\right).$$

Proof Let $\phi = \eta(\theta)$ and $\gamma(\phi) = d(\eta^{-1}(\phi))$. Then, the density/frequency admits the form

$$\exp\{\phi T(x) - \gamma(\phi) + S(x)\}, \qquad x \in \mathcal{X}, \phi \in \Phi,$$

and the conditions of Theorem 3.23 (p. 80) are all satisfied. Thus, the MLE $\hat{\phi}_n$ of $\phi_0 = \eta(\theta_0)$ is unique, and satisfies

$$\sqrt{n}(\hat{\phi}_n - \phi_0) \xrightarrow{d} N\left(0, \frac{1}{\gamma''(\phi_0)}\right),$$

where $0 < \frac{1}{\gamma''(\phi_0)} < \infty$.

It follows by the injective equivariance of maximum likelihood (Proposition 3.17, p. 72); see also Example 3.19 (p. 73)) that the unique MLE of θ_0 is $\hat{\theta}_n = \eta^{-1}(\hat{\phi}_n)$. The inverse function theorem now implies that $(\eta^{-1})'(y)$ exists in a small neighbourhood $B_\epsilon(\phi_0)$ of ϕ_0 and that

$$(\eta^{-1})'(\phi_0) = [\eta'(\eta^{-1}(\phi_0))]^{-1} = 1/\eta'(\theta_0).$$

Let $\eta^{-1}(\cdot)$ be equal to zero by convention outside of $B_\epsilon(\phi_0)$. Using the delta method (Theorem 2.27, p. 57), we obtain

$$\sqrt{n}(\hat{\theta}_n - \theta_0) = \sqrt{n}(\eta^{-1}(\hat{\phi}_n) - \eta^{-1}(\phi_0)) \xrightarrow{d} N\left(0, \frac{1}{(\eta'(\theta_0))^2\gamma''(\eta(\theta_0))}\right).$$

Note, however, that under the assumed conditions, we have shown in Exercise (23, p. 52) that

$$\frac{d''(\theta_0)\eta'(\theta_0) - d'(\theta_0)\eta''(\theta_0)}{[\eta'(\theta_0)]^3} = \text{Var}[T(X_i)] = \gamma''(\phi_0) > 0,$$

so that

$$\sqrt{n}(\hat{\theta}_n - \theta_0) \xrightarrow{d} N\left(0, \frac{[\eta'(\theta_0)]}{d''(\theta_0)\eta'(\theta_0) - d'(\theta_0)\eta''(\theta_0)}\right).$$

$\square$

▶ **Remark 3.28 (Asymptotic Variance and the Cramér–Rao Bound, Again)** Notice that for the usual parametrisation, we also have that the asymptotic mean of the MLE is equal to the true parameter, so that the asymptotic bias is again zero. Furthermore, if $\phi = \eta(\theta)$ and $\gamma(\phi) = d(\eta^{-1}(\phi))$, we note that

$$\mathbb{E}[(\ell'(\theta))^2] = \mathbb{E}\left[\left(\frac{\partial \ell(\theta)}{\partial \eta(\theta)} \frac{\partial \eta(\theta)}{\partial \theta}\right)^2\right] = (\eta'(\theta))^2 \mathbb{E}[(\ell'(\phi))^2] = (\eta'(\theta))^2 \mathrm{Var}[\tau(X_1, \ldots, X_n)]$$

$$= n(\eta'(\theta))^2 \frac{d''(\theta)\eta'(\theta) - d'(\theta)\eta''(\theta)}{[\eta'(\theta)]^3}$$

$$= n \frac{d''(\theta)\eta'(\theta) - d'(\theta)\eta''(\theta)}{[\eta'(\theta)]},$$

where we have used the same calculation as in Remark 3.25 (p. 82), and our result from Exercise 23 (p. 52). The inverse of the LHS is the Cramér–Rao lower bound (Theorem 3.9, p. 65). The inverse of the RHS is the asymptotic variance of $\hat{\theta}$. We thus see that the MLE of θ attains a performance that is close to optimal for n large.

▶ **Remark 3.29** A conclusion similar to that of Theorem 3.23 (p. 80) is in fact valid for a much broader class of distributions than just the exponential family. Under smoothness conditions on the density/frequency of the model, under analytical conditions enabling differentiation under the integral, and if the MLE $\hat{\theta}$ of θ is unique, one can show that

$$\sqrt{n}(\hat{\theta}_n - \theta_0) \xrightarrow{d} N\left(0, J^2(\theta_0)/I(\theta_0)\right),$$

where $I(\theta_0) = \mathbb{E}[(\ell'(\theta_0))^2]$ is the Fisher information and $J(\theta_0) = -\mathbb{E}[\ell''(\theta_0)]$. In fact, when we can differentiate under the integral, it is an easy exercise to show that $I(\theta) = J(\theta)$, and so the asymptotic variance becomes $1/I(\theta_0)$, attaining the Cramér-Rao bound.

Exercise 36 In the context of Corollary 3.27 (p. 83), prove that

$$\mathbb{E}\left[\frac{\partial}{\partial \theta} \log f(X_1, \ldots, X_n; \theta)\right] = 0, \quad \text{and}$$

$$\mathbb{E}\left[\left(\frac{\partial}{\partial \theta} \log f(X_1, \ldots, X_n; \theta)\right)^2\right] = -\mathbb{E}\left[\frac{\partial^2}{\partial \theta^2} \log f(X_1, \ldots, X_n; \theta)\right]. \quad (3.2)$$

Conclude that

$$I(\theta) = J(\theta) = \frac{d''(\theta)\eta'(\theta) - d'(\theta)\eta''(\theta)}{[\eta'(\theta)]}.$$

Exercise 37 Let $f(x; \theta)$ be a regular parametric model (not necessarily an exponential family) such that

$$\mathcal{X} = \{x \in \mathbb{R} : f(x; \theta) > 0\}$$

does not depend on θ, and f twice differentiable with respect to θ. Let $X_1, \ldots, X_n \overset{iid}{\sim} f(x; \theta)$. Show that the equality in 3.2 is equivalent to a regularity condition allowing us to interchange integration and differentiation. Reminder: for any $g : \mathbb{R}^n \to \mathbb{R}$,

$$\mathbb{E}[g(X)] = \int_{\mathcal{X}^n} g(x) f(x; \theta) \, dx \quad \text{when the integral exists} \quad (x = (x_1, \ldots, x_n) \in \mathbb{R}^n).$$

Exercise 38 We now consider two examples that fall outside of the realm of exponential families.
1. Let $X_1, \ldots, X_n \overset{iid}{\sim} Unif(0, \theta)$, where $\theta > 0$. Let $\widehat{\theta}_n$ be the MLE of θ. Find a sequence of real numbers a_n such that $a_n(\theta - \widehat{\theta}_n)$ converges in distribution to a non-degenerate limit (i.e. not a constant or infinity).
2. Consider $\widehat{\lambda}_n$, the estimator from exercise 33, p. 74. Find a sequence of real numbers a_n such that $a_n(\widehat{\lambda}_n - \lambda)$ converges in distribution to a non-degenerate limit.
    Hint : Show that if $X \sim Exp(\lambda)$, then $Y = aX \sim Exp(\lambda/a)$ for $a > 0$, then use Exercise 8 (p. 13).

3.3.4   Other Estimation Methods

In some situations, the MLE will not be determinable as an explicit function of the data. In these cases, one may need to numerically evaluate the MLE.

Example 3.30 (Cauchy MLE)

Suppose that $X_1, \ldots, X_n$ are iid random variables following the *Cauchy distribution* with density function

$$f(x; \theta) = \frac{1}{\pi(1 + (x - \theta)^2)}, \quad x \in \mathbb{R}.$$

The log likelihood function in this case is

$$\ell(\theta) = -\sum_{i=1}^{n} \log[1 + (X_i - \theta)^2] - n \log(\pi).$$

This is differentiable, and so if $\hat{\theta}$ is a maximum of $\ell(\theta)$, it must satisfy $\ell'(\hat{\theta}) = 0$, or equivalently

$$\sum_{i=1}^{n} \frac{2(X_i - \hat{\theta})}{1 + (X_i - \hat{\theta})^2} = 0.$$

The equation above cannot be explicitly solved to readily give us the form of the maximum likelihood estimator as an explicit function of the data. Therefore, the estimator remains implicitly defined. For any concrete sample $X_1 = x_1, \ldots, X_n = x_n$, we will need to solve the equation $\sum_{i=1}^{n} \frac{2(x_i - \hat{\theta})}{1 + (x_i - \hat{\theta})^2} = 0$ by some iterative/approximate solution method in order to get the numerical value of the maximum likelihood estimate. □

There are several numerical methods that one can employ in order to calculate the value of the maximum likelihood estimator in a specific sample (that is, in order to calculate the *estimate*). Among these, chief are the Newton–Raphson method, the method of bisection, the method of gradient descent and the EM-algorithm. Which one is most appropriate depends on the specific example. What is common to all of them is that they are iterative: they start at a given input value and iterate some operation until a convergence criterion is attained. Since the function ℓ' might not be monotone (and so may have multiple roots) it is important that the starting input value $\hat{\theta}_{(0)}$ be within a reasonable distance of the true maximum (for example, in Example 3.30, p. 86) we have a non-monotone ℓ'); otherwise, the algorithm may converge to a root that does not correspond to the maximum.

Example 3.31 (Newton–Raphson Iteration)

We consider the general idea behind the Newton–Raphson iteration. We wish to solve the equation $\ell'(\theta) = 0$, but cannot do so explicitly. Suppose that we somehow have a starting value $\hat{\theta}_{(0)}$ that is close to the true maximum $\hat{\theta}$. Since $\hat{\theta}$ is the overall maximum, it satisfies $\ell'(\hat{\theta}) = 0$. Now assume that ℓ is smooth enough that we can carry out a Taylor expansion. We will have (Theorem A.1, p. 159):

$$0 = \ell'(\hat{\theta}) = \ell'(\hat{\theta}_{(0)}) + (\hat{\theta} - \hat{\theta}_{(0)})\ell''(\hat{\theta}_{(0)}) + \frac{1}{2}(\hat{\theta} - \hat{\theta}_{(0)})^2 \ell'''(\theta_*),$$

where $\theta_* = \lambda\hat{\theta} + (1 - \lambda)\hat{\theta}_{(0)}$ for some $\lambda \in [0, 1]$. Now assuming that $|\hat{\theta} - \hat{\theta}_{(0)}|$ is small, the term $(\hat{\theta} - \hat{\theta}_{(0)})^2$ is negligible relative to the term $(\hat{\theta} - \hat{\theta}_{(0)})$. So, provided ℓ''' is bounded, we may write

$$\ell'(\hat{\theta}_{(0)}) + (\hat{\theta} - \hat{\theta}_{(0)})\ell''(\hat{\theta}_{(0)}) \simeq 0$$

which suggests

$$\hat{\theta} \simeq \hat{\theta}_{(0)} - \frac{\ell'(\hat{\theta}_{(0)})}{\ell''(\hat{\theta}_{(0)})}.$$

Now, the procedure can then be iterated by defining $\hat{\theta}_{(1)} = \hat{\theta}_{(0)} - \frac{\ell'(\hat{\theta}_{(0)})}{\ell''(\hat{\theta}_{(0)})}$, then $\hat{\theta}_{(2)} = \hat{\theta}_{(1)} - \frac{\ell'(\hat{\theta}_{(1)})}{\ell''(\hat{\theta}_{(1)})}$, and so on. This iteration will eventually lead to convergence. Guarantees on the convergence and how rapidly this will occur will depend on the specific form of ℓ.   □

How can we find a reasonable starting value $\hat{\theta}_{(0)}$? In some cases, reasonable starting values may be found by direct inspection.

Example 3.32 (Cauchy MLE, Continued)

Notice that the density $f(x;\theta)$ is symmetric about θ,

$$f(x;\theta) = \frac{1}{\pi(1 + (x - \theta)^2)}, \qquad x \in \mathbb{R}.$$

A potential starting value for θ is thus the median of $X_1, \ldots, X_n$. This could be employed as to initialise a Newton–Raphson iteration.   □

In other cases, things may not be as clear.

Example 3.33 (Gamma MLE)

Let $X_1, \ldots, X_n \overset{iid}{\sim}$ Gamma$(r, 1)$ and suppose we wish to estimate the parameter r by the method of maximum likelihood. The likelihood is

$$L(r) = \prod_{i=1}^{n} \frac{1}{\Gamma(r)} X_i^{r-1} e^{-X_i},$$

with corresponding loglikelihood

$$\ell(r) = -n \log \Gamma(r) + (r - 1) \sum_{i=1}^{n} \log X_i - \sum_{i=1}^{n} X_i.$$

Differentiating and setting the loglikelihood equal to zero, we find that the MLE $\hat{r}$ must satisfy,

$$\frac{\Gamma'(\hat{r})}{\Gamma(\hat{r})} = \overline{\log X} = \frac{1}{n} \sum_{i=1}^{n} \log X_i.$$

This equation cannot be solved explicitly. Worse, even, there is no immediate plausible value for r by simple inspection of the form of the density. In this case, we need some other way of determining a starting value for a Newton–Raphson iteration.   □

To address the issue of finding general methods for the determination of starting values $\hat{\theta}_{(0)}$, it is useful to have estimation methods that will yield some explicit estimates that could be used to initialise iterative techniques in search of a maximum likelihood estimate. These methods do not necessarily need to be as efficient as the method of maximum likelihood, but at least produce estimators that are reasonably good. A widely used such method is the *method of moments*.

3.3.4.1 The Method of Moments

Consider first a one-dimensional problem, where $\{ f_\theta : \theta \in \Theta \}$ is a one-parameter regular model, $\Theta \subseteq \mathbb{R}$, and $X_1, .., X_n$ is an iid sample generated by a true parameter $\theta \in \Theta$. The method of moments is motivated by the following heuristic. Assuming $\mathbb{E}[|X_1|] < \infty$, the law of large numbers tells us that

$$\frac{1}{n} \sum_{i=1}^{n} X_i \xrightarrow{p} \mathbb{E}[X_1].$$

But $\mathbb{E}[X_1] = \int_{-\infty}^{+\infty} x f(x; \theta) dx$ depends on the unknown parameter θ, so we can write that $\mathbb{E}[X_1] = m(\theta)$ for some m. Rephrasing, we have

$$\frac{1}{n} \sum_{i=1}^{n} X_i \xrightarrow{p} m(\theta)$$

and so, in other words, we expect that, for n large enough

$$\frac{1}{n} \sum_{i=1}^{n} X_i \simeq m(\theta)$$

for θ the true parameter. So, if $\hat{\theta}$ is to be close to θ, we expect that it should satisfy

$$\frac{1}{n} \sum_{i=1}^{n} X_i \simeq m(\hat{\theta}).$$

This motivated the method of moments:

Definition 3.34 (Method of Moments Estimator: Single Parameter Case)

Let $X_1, \ldots, X_n$ be an iid random sample from a distribution F_θ with density (or mass function) $f(x; \theta)$. Assume that $\mathbb{E}|X_1| < \infty$ for all $\theta \in \Theta \subseteq \mathbb{R}$. Let $\hat{\theta}$ be such that

$$\frac{1}{n} \sum_{i=1}^{n} X_i = m(\hat{\theta}),$$

where

$$m(\theta) = \int_{-\infty}^{+\infty} x f(x; \theta) dx, \quad \theta \in \mathbb{R}.$$

Then $\hat{\theta}$ is called a *Method of Moments* (MoM) estimator of θ.

In other words, the method of moments says that we should equate the theoretical first moment with the observed empirical first moment. This will yield an equation with the unknown being the parameter we wish to estimate. Solving this equation for the unknown will yield an estimator of θ, the *Method of Moments* estimator. The thing to note here is that this equation is typically easier to solve than an MLE score equation, because the data have been separated on one side of the equation (yielding a single numerical constant given the observed sample) and the function of the parameter on the other side. So rather than having an equation of the form

$$g(X_1, \ldots, X_n, \theta) = 0$$

we have the easier problem of the form

$$g(\theta) = h(X_1, \ldots, X_n).$$

Here is an illustration of the technique in a simple example.

Example 3.35 (Uniform MoM Estimator)

Let $X_1, \ldots, X_n \overset{iid}{\sim} Unif(0, \theta)$, and suppose that we wish to estimate $\theta \in \mathbb{R}_+$. In this case we have a single parameter, so that the MoM estimator of θ, say $\hat{\theta}$ must be such that

$$\frac{1}{n} \sum_{i=1}^{n} X_i = m(\hat{\theta}).$$

In this case,

$$m(\theta) = \int_0^\theta \frac{x}{\theta} dx = \frac{\theta}{2}.$$

Therefore, the method of moments estimator is

$$\hat{\theta} = \frac{2}{n} \sum_{i=1}^{n} X_i.$$

$\square$

In case we need to estimate multiple parameters, say $\theta = (\theta_1, \ldots, \theta_p)^\top$, then the method of moments asks that we equate the first p empirical moments to the first p theoretical moments and obtain a system of p equations with the p parameters as unknowns. Solving this system will yield an estimator for θ.

Definition 3.36 (Method of Moments Estimator: Multiparameter Case)

Let $X_1, \ldots, X_n$ be an iid random sample from a distribution F_θ with density (or mass function) $f(x; \theta)$. Assume that $\mathbb{E}|X_1|^p < \infty$ for all $\theta \in \Theta \subseteq \mathbb{R}^p$. Let $\hat{\theta}$ be such that

$$\frac{1}{n} \sum_{i=1}^n X_i^k = m_k(\hat{\theta}), \quad k = 1, \ldots, p$$

where

$$m_k(\theta) = \int_{-\infty}^{+\infty} x^k f(x; \theta) dx, \quad \theta \in \mathbb{R}^p, \ k = 1, \ldots, p.$$

Then $\hat{\theta}$ is called a *Method of Moments* (MoM) estimator of θ.

The following example illustrates a two-parameter situation where the Method of Maximum likelihood does not yield explicit estimators, but the Method of Moments does.

Example 3.37 (Gamma MoM Estimator)

Suppose that $X_1, \ldots, X_n \overset{iid}{\sim} \text{Gamma}(r, \lambda)$ and we wish to estimate the parameter vector $(r, \lambda)^\top$. The first two moment equations are

$$\frac{1}{n} \sum_{i=1}^n X_i = m_1(\hat{r}, \hat{\lambda}) \quad \text{and} \quad \frac{1}{n} \sum_{i=1}^n X_i^2 = m_2(\hat{r}, \hat{\lambda}).$$

But we have seen that

$$m_1(r, \lambda) = r/\lambda \quad \text{and} \quad m_2(r, \lambda) = \mathbb{E}^2[X_1] + \text{Var}[X_1] = r^2/\lambda^2 + r/\lambda^2 = r(r+1)/\lambda^2.$$

Solving the system of moment equations with respect to the unknown parameter yields the estimates

$$\hat{r} = \frac{n\bar{X}^2}{\sum_{i=1}^n (X_i - \bar{X})^2} \quad \text{and} \quad \hat{\lambda} = \frac{n\bar{X}}{\sum_{i=1}^n (X_i - \bar{X})^2}.$$

$\square$

Exercise 39   Let $X_1, \ldots, X_n$ be an iid sample from the density

$$f(x; \theta) = \begin{cases} 3\theta^3 x^{-4}, & \text{if } x \geq \theta, \\ 0, & \text{otherwise,} \end{cases}$$

where $\theta > 0$.
1. Find the method of moment estimator $\hat{\theta}_n^{\text{MoM}}$ of θ.
2. Find the ML estimator $\hat{\theta}_n^{\text{MV}}$ of θ.
3. Show that $\hat{\theta}_n^{\text{MoM}}$ is unbiased, but $\hat{\theta}_n^{\text{MV}}$ is biased.
4. Calculate the mean squared error of $\hat{\theta}_n^{\text{MoM}}$ and $\hat{\theta}_n^{\text{MV}}$. Which estimator would one prefer?

A drawback of the Method of Moments is that it is not guaranteed to always work. To even be able to define the procedure, we require the existence of a p-th absolute moment in order for the method to work in a p-parameter problem. If such a moment does not exist, the method fails.

Example 3.38 (MoM Failure in Cauchy Case)

Let $X_1, \ldots, X_n$ be iid random variables following the *Cauchy distribution* with density function

$$f(x; \theta) = \frac{1}{\pi(1 + (x - \theta)^2)}, \qquad x \in \mathbb{R}.$$

Notice that

$$m_1(0) = \frac{1}{\pi} \int_{-\infty}^{+\infty} \frac{x}{1 + x^2} dx = \frac{1}{\pi} \int_{-\infty}^{0} \frac{x}{1 + x^2} dx + \frac{1}{\pi} \int_0^{+\infty} \frac{x}{1 + x^2} dx = -\infty + \infty \quad \text{(undefined)}$$

Therefore, the moment equations are undefined, and no MoM exists. □

In general, when the moment generating function exists, then the method of moment is well defined, regardless of the dimension of the parameter. Still, there can be no guarantees that the system of equations produced will always have a solution. We will not further pursue conditions under which this could be enforced to be the case.

3.4   Estimation Methods vs Estimators vs Estimates

We conclude this chapter by a short remark on terminology, that can sometimes be the source of some confusion. Specifically we distinguish between the notions of an *estimation method*, *estimator* and an *estimate*. Here are some points to bear in mind:
1. An *estimation method* is a general principle or procedure that can be applied in any particular parametric model in order to obtain *estimators*. We saw examples of how we can apply the method of maximum likelihood to get estimators of parameters in the Bernoulli, exponential, normal and uniform distributions.
2. It can very well happen that the same *estimation method* produces different *estimators* when applied to two different parametric models. For example, the

method of maximum likelihood produces the estimator $\overline{X}$ for the mean of a normal distribution, and the estimator $1/\overline{X}$ for the mean of an exponential distribution.

3. It can also happen that two different *estimation methods* produce the same *estimator* in the same model. For example, the maximum likelihood estimator for the mean of a normal distribution coincides with the method of moment estimator for the mean of a normal distribution.

4. An *estimate* is the specific value that an *estimator* takes when evaluated on the basis of an observed sample. Remember: an *estimator* is a random variable. The realisation of this random variable is called an *estimate*.

Tests of Hypotheses for Model Parameters

4

So far, we have considered the problem of point estimation: given a parametric model $\{F_\theta : \theta \in \Theta\}$, and an iid sample $X_1, \ldots, X_n$ issued from some specific F_θ, estimate the value of θ that generated the sample. There are many contexts, however, where the precise value of the true parameter is not the primary object of our interest. Rather, we are more interested in using the sample to ascertain whether the true value of the parameter belongs to some specific subset of parameter values or not.

Example 4.1 (Coin Tossing)

For a simple example, consider a situation where we wish to ascertain whether a coin is fair, or is biased. We may flip the coin n times and record the outcome of each coin toss. We then wish to use the outcomes in order to decide whether the probability of heads is equal to 1/2 or whether it is different from 1/2. We could formalise this problem by saying that we have $X_1, \ldots, X_n \overset{iid}{\sim} Bern(p)$ and wish to decide whether $p \in \{\frac{1}{2}\}$ or $p \in (0, 1) \setminus \{\frac{1}{2}\}$. $\qquad\square$

To make things more concrete, suppose that we know that the parameter has to lie in one of two sets: either in Θ_0 or in Θ_1, where $\Theta_0 \cap \Theta_1 = \emptyset$. We wish to employ the sample $X_1, .., X_n$ that we have at our disposal in order to decide which is the case. This setup arises very often in the sciences, where there are two competing scientific hypotheses. The *null hypothesis* H_0, that states that $\theta \in \Theta_0$,

$$H_0 : \theta \in \Theta_0$$

and the competing *alternative hypothesis* that instead postulates that $\theta \in \Theta_1$,

$$H_1 : \theta \in \Theta_1.$$

© Springer International Publishing Switzerland 2016
V.M. Panaretos, *Statistics for Mathematicians*, Compact Textbooks in Mathematics,
DOI 10.1007/978-3-319-28341-8_4

Example 4.2 (Search for the Higgs Boson)

One of the biggest questions in particle physics in the last quarter century was whether or not the infamous *Higgs boson* exists or not. One way to detect whether this elementary particle indeed exists is via its decay into two photons. Using the Standard Model of particle physics, we can compute how many such diphoton events would be produced on average if there was no Higgs boson. Let's denote this by b. Similarly, we can also compute how many extra diphotons would be produced on average if the Higgs particle did indeed exist. Let's call this s. Observed diphoton events are well documented to follow the Poisson distribution with some mean, say μ. Therefore, the null hypothesis corresponding to the state of nature if the Higgs boson did not exist would be

$$H_0 : \mu = b$$

and the competing *alternative hypothesis* (describing the state of nature if the Higgs boson existed),

$$H_1 : \mu = b + s.$$

$\square$

The statistical problem of hypothesis testing considers how to efficiently employ the sample in order to decide between the two competing hypotheses H_0 and H_1. To do this, we must first consider *how* one can employ the sample to this aim, and what sorts of error one can incur as a result. The next section introduces the relevant notions

4.1   Test Functions and Error Types

The decision between H_0 and H_1 is to be made on the basis of the observed sample $X_1, .., X_n$. A simple way to state this mathematically is via the following definition.

Definition 4.3 (Test Function)
A test function δ is any function $\delta : \mathcal{X}^n \to \{0, 1\}$.

A test function takes the value '0' when we rule in favour of H_0 based on the sample, and it takes the value '1' when we rule in favour of H_1. A test function will typically take the value 0 or 1 depending on whether or not the sample satisfies a certain condition. In other words, test functions are usually constructed by

$$\delta(X_1, \ldots, X_n) = \begin{cases} 1, & \text{if } T(X_1, \ldots, X_n) \in C, \\ 0, & \text{if } T(X_1, \ldots, X_n) \notin C. \end{cases}$$

where T is a statistic called a *test statistic* and C a set in the range of T called the *critical region*. Notice that in compact notation, we may write

$$\delta(X_1, \ldots, X_n) = \mathbf{1}\{T(X_1, \ldots, X_n) \in C\}.$$

Therefore the choice of the test function rests on the choice of T and of C. How should we make this choice in order to obtain a good test function? Notice that δ is always a Bernoulli random variable, since it takes the values 0 and 1,

$$\delta = \begin{cases} 1, & \text{with probability } \mathbb{P}[T(X_1, \ldots, X_n) \in C], \\ 0, & \text{with probability } \mathbb{P}[T(X_1, \ldots, X_n) \notin C]. \end{cases}$$

This Bernoulli variable may give different decisions for different realisations of the random sample. Therefore, just as with the problem of point estimation (where we needed to choose good estimators), our choice of a test function must be guided by a careful consideration of what types of errors one can commit. A good test function will then be a δ whose sampling behaviour fares well relative to these error criteria.

In hypothesis testing, there are two possible states of nature, and two possible decisions that we can make. Therefore, the "error landscape" is described by the following table:

Decision/Truth	H_0	H_1
0	No error	Type II error
1	Type I error	No error

When the truth is $H_0 : \theta \in \Theta_0$, we hope that the distribution of $\delta(X_1, \ldots, X_n)$ will concentrate around the value 0. Conversely, when the truth is $H_1 : \theta \in \Theta_1$, we hope that the distribution of $\delta(X_1, \ldots, X_n)$ will concentrate around the value 1. Therefore, a good decision rule should concentrate around the value i, whenever H_i is true, for $i \in \{0, 1\}$. So, by a slight abuse of terminology, we can compare decision rules δ by looking at something like their "mean square error",

$$MSE(\delta, H_i) = \mathbb{E}_\theta[(\delta - i)^2], \qquad i \in \{0, 1\}.$$

Since δ is a Bernoulli variable and i takes values in $\{0, 1\}$, we have

$$MSE(\delta, H_i) = \begin{cases} \mathbb{P}_\theta[\delta = 1], & \text{if } \theta \in \Theta_0, \\ \mathbb{P}_\theta[\delta = 0], & \text{if } \theta \in \Theta_1. \end{cases}$$

This motivates the following definition.

Definition 4.4 (Error Probabilities)

Let $H_0 : \theta \in \Theta_0$ and $H_1 : \theta \in \Theta_1$ be two competing hypotheses. The Type I error Probability is defined to be the mapping $h : \Theta_0 \to [0, 1]$,

$$h(\theta) = \mathbb{P}_\theta[\delta = 1], \qquad \theta \in \Theta_0.$$

The Type II error Probability is defined to be the mapping $g : \Theta_1 \to [0, 1]$,

$$g(\theta) = \mathbb{P}_\theta[\delta = 0], \qquad \theta \in \Theta_1.$$

▶ **Remark 4.5** That the two error probabilities are functions of θ simply reflects the fact that our error depends on the true state of nature: for some θ it will be easier to distinguish between Θ_0 and Θ_1 than for some others. For example, consider $\Theta_0 = (-\infty, b]$ and $\Theta_1 = (b, \infty)$. For a given test function δ, we expect that it will be easier to get things right when the true parameter is away from the boundary value b, then when the true parameter is close to b.

▶ **Remark 4.6 (Warning on Error Probabilities)**   Notice that $h(\theta) \neq 1 - g(\theta)$ since the two functions are defined over different domains. It is a common mistake to not realise this.

In order to have a good test function, we must try to choose the test statistic T and the critical region C in such a way that the *probability of type I error* be small for all $\theta \in \Theta_0$ and at the same time the *probability of type II* error be small for all values of $\theta \in \Theta_1$. The *Neyman–Pearson* framework in the next paragraph considers how to attack this problem.

▶ **Remark 4.7 (Type I vs Type II Error)**   It is no coincidence that the two types of error are given two different names, and in fact names that suggest that one kind of error is of primary importance (type I) and the other is secondary (type II). In many practical contexts, the two hypotheses are asymmetric: making one kind of error is far more serious than the other type of error. The more serious type of error is named the Type I error and the other is the Type II error. Therefore, in all practical situations, H_0 is chosen to be the hypothesis whose false rejection is more harmful.

Example 4.8 (Spam Filter)

Suppose we wish an automatic test function to decide whether a new email is spam or not. The new message contains n words $X_1, .., X_n$ and we need a test function in order to decide between two competing hypotheses: "spam" versus "not spam". Notice that marking a message as spam when it is in fact not can have serious consequences (since we will not see it and it could be important). Marking a message as "not spam" when in fact it is spam is annoying, but perhaps not as big of a problem. In this context, it is reasonable to define "H_0 : Message is not spam" and "H_1 : Message is spam". If we do so, the type I error will be precisely the probability to mark a message as spam when it is not.   □

Exercise 40   Consider the following statistical hypothesis testing scenarios. Write down in each case the two competing hypotheses and the two types of errors you can make. Based on this, decide which hypothesis should be the null hypothesis H_0 and which one the alternative hypothesis H_1.

1. You are a physicist working on an experiment to detect dark matter particles. Test if your data contains a dark matter signal.
2. You are trying to decide if you can drive home after attending a wine tasting. Test if your blood alcohol level is above the legal limit for driving.
3. Barack Obama and Mitt Romney were the leading candidates in the 2012 US presidential elections. You are the campaign manager for Mr. Obama trying to decide how to best allocate his campaign funds. Test whether Obama is leading the race in the state of Iowa. How would your test change if you were the campaign manager for Mr. Romney?
4. You are a scientist working at a pharmaceutical company. You have developed a new drug for reducing high blood pressure. Test if your drug works as advertised.

Exercise 41 Let $X_1, \ldots, X_n$ be an iid sample from an $N(\mu, 1)$ distribution. We will test $H_0 : \mu = 0$ against the alternative $H_1 : \mu \neq 0$ using the test statistic

$$T_n(X_1, \ldots, X_n) = \bar{X}_n = \frac{1}{n} \sum_{i=1}^{n} X_i,$$

and corresponding test function

$$\delta(X_1, \ldots, X_n) = \begin{cases} 1, & \text{if } |T_n(X_1, \ldots, X_n)| \geq Q, \\ 0, & \text{otherwise}, \end{cases}$$

where $Q > 0$.
1. Find the probability of committing a type I error.
2. Find the probability of committing a type II error.
3. How do these vary as we increase Q?

Exercise 42 Let $X_1, \ldots, X_n$ be an iid sample from the Bernoulli(p) distribution, with $p \in (0, 1)$. We will test the null hypothesis $H_0 : p = \frac{1}{2}$ against the alternative $H_1 : p \in (0, 1) \setminus \{1/2\}$ using the test statistic

$$T_n(X_1, \ldots, X_n) = \bar{X}_n - \frac{1}{2} = \frac{1}{n} \sum_{i=1}^{n} X_i - \frac{1}{2},$$

and the corresponding test function

$$\delta(X_1, \ldots, X_n) = \begin{cases} 1, & \text{if } |T_n(X_1, \ldots, X_n)| \geq Q, \\ 0, & \text{otherwise}, \end{cases}$$

where $Q \in (0, \frac{1}{2}]$.

1. Find the probability of committing a type I error.
2. Find the probability of committing a type II error.
3. How do these vary as we increase Q?

Exercise 43 (Bonferroni Correction, Multiple Testing) For each $j = 1, \ldots, J$, let

$$\{X_{1j}, \ldots, X_{nj}\}$$

be iid Bernoulli random variables with (unknown) success probability $p_j \in (0, 1)$, and $n > 1$. Note that the variables are independent for fixed j and varying i, but could dependent for fixed i and varying j (e.g. X_{ij} may be a yes/no answer of the ith individual to the jth question of a customer survey). We wish to test the hypothesis pair:

$$\left\{ \begin{array}{ll} H_0 : p_j \geq \frac{1}{2} & \forall\, j = 1 \ldots, J \\ H_1 : \exists\, j \in \{1, \ldots, J\} : & p_j < \frac{1}{2} \end{array} \right\}.$$

(in our example: are customers on average satisfied on all J issues, or do there exist issues where customers are dissatisfied, on average?). Construct a test for the hypothesis pair that respects a given level $\alpha \in (0, 1)$.

4.2   The Neyman–Pearson Framework

Recall the closing of the previous paragraph: we must try to choose the test statistic T and the critical region C in such a way that the *probability of type I error* be small for all $\theta \in \Theta_0$ and at the same time the *probability of type II* error be small for all values of $\theta \in \Theta_1$. Is it possible to always make both these probabilities small, for all θ in the respective sets Θ_0 and Θ_1?

Unfortunately, the answer is **no**. Here is why. Let $\delta(X_1, \ldots, X_n) = \mathbf{1}\{T(X_1, \ldots, X_n) \in C\}$ be a test function, and suppose that we wish to reduce its type I error probability,

$$h(\theta) = \mathbb{P}_\theta[\delta = 1], \qquad \theta \in \Theta_0.$$

over all $\theta \in \Theta_0$. To do this, we must "reject less often", that is, we must replace C by a set $C_* \subset C$ and obtain the new test function $\delta_* = \mathbf{1}\{T(X_1, \ldots, X_n) \in C_*\}$. Observe that

$$\mathbb{P}_\theta[\delta_* = 1] = \mathbb{P}[T(X_1, \ldots, X_n) \in C_*]$$
$$\leq \mathbb{P}[T(X_1, \ldots, X_n) \in C] = \mathbb{P}_\theta[\delta = 1], \qquad \forall\, \theta \in \Theta_0.$$

But now notice that $C_* \subset C \implies C_*^c \supset C^c$ and so

$$\mathbb{P}_\theta[\delta_* = 0] = \mathbb{P}[T(X_1, \ldots, X_n) \notin C_*]$$
$$\geq \mathbb{P}[T(X_1, \ldots, X_n) \notin C] = \mathbb{P}_\theta[\delta = 0], \qquad \forall \, \theta \in \Theta_1.$$

In other words, by trying to reduce the type I error, we have increased the type II error! By symmetry, we can also show that a similar attempt to reduce the type II error would inflate the type I error (for two concrete examples, consider Exercises 41 and 42, p. 99).

It seems that we cannot insist on simultaneously reducing the two types of errors, and we need to make some concessions. The fundamental premise of the *Neyman–Pearson Framework* is that since type I error is more important, we should first try to fix the corresponding probability of type I error to some low level. Once this is fixed, we can then shift focus on getting a low type II error probability. We describe the framework in the following steps:

Definition 4.9 (Neyman–Pearson Framework)

Let $H_0 : \theta \in \Theta_0$ and $H_1 : \theta \in \Theta_1$ be two competing hypotheses.

1. Fix an $\alpha \in (0, 1)$ and call it the *significance level* or just *level* of the test.
2. Consider only test functions $\delta : \mathcal{X}^n \to \{0, 1\}$ that respect the level, i.e. test functions δ such that

$$\sup_{\theta \in \Theta_0} \mathbb{P}_\theta[\delta = 1] \leq \alpha.$$

For ease of reference, we call this class $\mathcal{D}(\Theta_0, \alpha)$. In other words,

$$\mathcal{D}(\Theta_0, \alpha) = \left\{ \delta : \mathcal{X}^n \to \{0, 1\} \,\middle|\, \sup_{\theta \in \Theta_0} \mathbb{P}_\theta[\delta = 1] \leq \alpha \right\}.$$

3. Within the class of test functions $\mathcal{D}(\Theta_0, \alpha)$, compare test functions by considering which has lower type II error probability

$$g(\theta) = \mathbb{P}_\theta[\delta = 0], \qquad \theta \in \Theta_1.$$

Equivalently, one can compare test functions by considering which has higher *power*

$$\beta(\theta) = 1 - g(\theta) = \mathbb{P}_\theta[\delta = 1], \qquad \theta \in \Theta_1.$$

The intuition behind the Neyman–Pearson reasoning is as follows: we know that committing a type I error is most harmful. Therefore, we must make it our top priority to tightly control the probability of type I error. For this reason, we must only consider test functions whose type I error probability never exceeds some level

α (usually taken to be small, e.g. $\alpha = 0.05$). Given that this restriction is satisfied, we can then turn to trying to minimise the type II error probability, or equivalently to maximising the power.

Exercise 44
1. In the context of exercise 41, p. 99, find the smallest value of Q for which the significance level is equal to some $\alpha \in (0, 1)$. Evaluate this for $\alpha = 0.05$ and $n = 10$. Find the supremum (over the parameter space) of the probability of type II error for that value of Q.
2. In the context of exercise 42, p. 99, suppose that $n = 10$. Find the values of Q for which the significance level is $\alpha = 0.05$. What is different here as opposed to the first part of the exercise? Why?

4.3   Methods for Constructing Test Functions

Now we know what a test function is, what sorts of error we can expect to incur, and what properties test functions should satisfy (as dictated by the Neyman–Pearson framework). So it's time to turn to the question of finding general methods for constructing test functions. It turns out that how one constructs a test function can depend very heavily on the types of hypotheses under consideration. To simplify things, we will consider only 1-dimensional parameters θ, and hypothesis pairs of the form:
1. **Simple vs Simple** ($H_0 : \theta = \theta_0$, $H_1 : \theta = \theta_1$, for some given $\theta_0 \neq \theta_1$).
2. **Left Unilateral vs Right Unilateral**: ($H_0 : \theta \leq \theta_0$, $H_1 : \theta > \theta_0$, for some given θ_0).
3. **Right Unilateral vs Left Unilateral**. ($H_0 : \theta \geq \theta_0$, $H_1 : \theta < \theta_0$, for some given θ_0).
4. **Simple vs Bilateral**: ($H_0 : \theta = \theta_0$, $H_1 : \theta \neq \theta_0$, for some given θ_0).
In short, we will only consider pairs of the form:

$$\underbrace{\left\{ \begin{matrix} H_0 : \theta = \theta_0 \\ H_1 : \theta = \theta_1 \end{matrix} \right\}}_{\text{simple vs simple}} \text{ or } \underbrace{\left\{ \begin{matrix} H_0 : \theta \leq \theta_0 \\ H_1 : \theta > \theta_0 \end{matrix} \right\} \text{ or } \left\{ \begin{matrix} H_0 : \theta \geq \theta_0 \\ H_1 : \theta < \theta_0 \end{matrix} \right\}}_{\text{unilateral vs unilateral}} \text{ or } \underbrace{\left\{ \begin{matrix} H_0 : \theta = \theta_0 \\ H_1 : \theta \neq \theta_0 \end{matrix} \right\}}_{\text{simple vs bilateral}}$$

While this may seem restrictive, it encompasses a large variety of applied situations. In applications, it is often sought to decide between two parameter values, or to decide whether a certain parameter is above, below or just deviates from a given threshold.

Now, before we proceed with considering methods for constructing tests, we recall that in the Neyman–Pearson framework (Definition 4.9, p. 101) we set a level α, and consider only test functions that respect this level. In other words, we restrict attention to elements in $\mathcal{D}(\Theta_0, \alpha)$. Within this class, we compare test functions by comparing their corresponding power functions. This motivates the following definition of optimality:

Definition 4.10 (Optimal Tests)

A test function δ of $H_0 : \theta \in \Theta_0$ vs $H_1 : \theta \in \Theta_1$ is called optimal at level α (or uniformly most powerful at level α) if the following two hold.

1. $\delta \in \mathcal{D}(\Theta_0, \alpha)$.
2. $\mathbb{P}_{\theta_1}[\psi = 1] \leq \mathbb{P}_{\theta_1}[\delta = 1]$ for all $\theta_1 \in \Theta_1$ and all $\psi \in \mathcal{D}(\Theta_0, \alpha)$.

Therefore, we wish to find methods that yield tests respecting the level, and with as high power as possible, for as many elements in the alternative set Θ_1 as possible. As it turns out, sometimes there do exist testing methods that are optimal—when this is the case, there is no reason to consider any other method. The existence of such optimal tests, though, depends strongly on the structure of Θ_0 and Θ_1, and also on the particular probability model under study. We will therefore structure our study of testing methods according to the types of pairs considered. Here is an overview:

(a) **Simple vs Simple**: In this case we will be able to find optimal tests that remain optimal regardless of the underlying model.

(b) **Unilateral**: In this case we will be able to find optimal tests for specific classes of models, specifically for the exponential family of distributions.

(c) **Bilateral**. In this case we will demonstrate that no optimal tests exist in general. We will nevertheless propose two general methods inspired by the concept of likelihood, that perform well in general.

4.3.1   Simple Case

In the case of a simple vs a simple hypothesis, the following result due to Neyman and Pearson gives us a method for constructing optimal tests.

Lemma 4.11 (Neyman–Pearson) *Let* $X = (X_1, \ldots, X_n)$ *have joint density (or frequency) function* $f_X(x; \theta)$ *and suppose we wish to test*

$$H_0 : \theta = \theta_0 \qquad vs \qquad H_1 : \theta = \theta_1.$$

at some level $\alpha \in (0, 1)$, *for* $\theta_0 \neq \theta_1$. *If the random variable*

$$\Lambda(X) = \frac{f_X(X_1, \ldots, X_n; \theta_1)}{f_X(X_1, \ldots, X_n; \theta_0)} = \frac{L(\theta_1)}{L(\theta_0)}$$

is such that there exists a $Q > 0$ *satisfying*

$$\mathbb{P}_{\theta_0}[\Lambda > Q] = \alpha$$

then the test whose test function is given by

$$\delta(X) = 1\{\Lambda(X) > Q\},$$

is an optimal (most powerful) *test of H_0 versus H_1 at significance level α.*

▶ **Remark 4.12** A sufficient condition for the existence of a suitable Q for any $\alpha \in (0, 1)$ is that Λ be a continuous random variable under the null hypothesis. If the distribution of Λ under H_0 is discrete or has discontinuities, there may exist $\alpha \in (0, 1)$ such that $\mathbb{P}_{\theta_0}[\Lambda > Q] = \alpha$ cannot be satisfied for any $Q > 0$.

Notice the intuition behind the test: we know that the method of maximum likelihood is a very good estimation method. The higher the likelihood of a parameter, the more plausible this parameter value is as a guess for the true parameter. So, in order to test $H_0 : \theta = \theta_0$ against $H_1 : \theta = \theta_1$, we decide to compare the value of the likelihood function at the two competing parameter values θ_0 and θ_1. If the likelihood of θ_1 is *significantly* higher than the likelihood of θ_0, then we reject H_0 in favour of H_1. How much higher qualifies as *significantly higher*? The theorem tells us that Q-times higher is significantly higher, where Q is a critical value chosen so that the level α be respected.

Proof of Lemma 4.11 We need to verify properties (1) and (2) in Definition 4.10 (p. 103). Since Q is such that $\mathbb{P}_{\theta_0}[\Lambda > Q] = \alpha$, then we immediately have that

$$\mathbb{P}_{\theta_0}[\delta = 1] = \alpha \qquad (\text{since } \mathbb{P}_{\theta_0}[\delta = 1] = \mathbb{P}_{\theta_0}[\Lambda > Q]). \qquad (4.1)$$

Therefore $\delta \in \mathcal{D}(\{\theta_0\}, \alpha)$ (i.e. δ indeed respects the level α) which yields (1).
To show (2), let $\psi \in \mathcal{D}(\{\theta_0\}, \alpha)$. For notational ease, write $(X_1, \ldots, X_n)^\top = X$ and $(x_1, \ldots, x_n)^\top = x$. Without loss of generality assume that f_X is a density function (otherwise replace any integrals that follow by sums), and observe that

$$f(x; \theta_1) - Q \cdot f(x; \theta_0) > 0 \text{ if } \delta(x) = 1 \quad \& \quad f(x; \theta_1) - Q \cdot f(x; \theta_0) \leq 0 \text{ if } \delta(x) = 0.$$

Therefore, since ψ can only take the values 0 or 1,

$$\psi(x)(f(x; \theta_1) - Q \cdot f(x; \theta_0)) \leq \delta(x)(f(x; \theta_1) - Q \cdot f(x; \theta_0))$$

$$\int_{\mathcal{X}^n} \psi(x)(f(x; \theta_1) - Q \cdot f(x; \theta_0)) dx \leq \int_{\mathcal{X}^n} \delta(x)(f(x; \theta_1) - Q \cdot f(x; \theta_0)) dx$$

Rearranging the terms yields

$$\int_{\mathcal{X}^n} (\psi(x) - \delta(x)) f(x; \theta_1) dx \le Q \int_{\mathcal{X}^n} (\psi(x) - \delta(x)) f(x; \theta_0) dx$$

$$\implies \mathbb{E}_{\theta_1}[\psi(X)] - \mathbb{E}_{\theta_1}[\delta(X)] \le Q \left(\mathbb{E}_{\theta_0}[\psi(X)] - \mathbb{E}_{\theta_0}[\delta(X)]\right)$$

$$\implies \mathbb{P}_{\theta_1}[\psi(X) = 1] - \mathbb{P}_{\theta_1}[\delta(X) = 1] \le Q \left(\mathbb{P}_{\theta_0}[\psi(X) = 1] - \mathbb{P}_{\theta_0}[\delta(X) = 1]\right)$$

Equation (4.1), combined with the fact that $\psi \in \mathcal{D}(\{\theta_0\}, \alpha)$ and $Q > 0$, implies that the right-hand side is non-positive. This proves (2) in Definition 4.10 (p. 103), and thus completes the proof.   □

Example 4.13

Let $X_1, \dots, X_n \overset{iid}{\sim} \mathrm{Exp}(\lambda)$ and let $\lambda_1 > \lambda_0$ be two constants. Consider the problem of testing the hypothesis pair:

$$\begin{cases} H_0 : & \lambda = \lambda_0 \\ H_1 : & \lambda = \lambda_1. \end{cases}$$

The likelihood is

$$f(X_1, \dots, X_n; \lambda) = \prod_{i=1}^{n} \lambda e^{-\lambda X_i} = \lambda^n e^{-\lambda \sum_{i=1}^{n} X_i}$$

So according to the Neyman–Pearson lemma we must base our test on the statistic

$$\Lambda(X_1, \dots, X_n) = \frac{f(X_1, \dots, X_n; \lambda_1)}{f(X_1, \dots, X_n; \lambda_0)} = \left(\frac{\lambda_1}{\lambda_0}\right)^n \exp\left[(\lambda_0 - \lambda_1) \sum_{i=1}^{n} X_i\right].$$

rejecting the null if $\Lambda \ge Q$, for Q such that $\mathbb{P}_{\lambda_0}[\Lambda(X_1, \dots, X_n) \ge Q] = \alpha$, provided such a Q exists. To determine whether it does exist, and if so what it is, we note that $\Lambda(X_1, \dots, X_n)$ is a decreasing function of $\tau(X_1, \dots, X_n) = \sum_{i=1}^{n} X_i$ (since $\lambda_0 < \lambda_1$). Therefore

$$\Lambda(X_1, \dots, X_n) \ge Q \iff \tau(X_1, \dots, X_n) \le q$$

for some q, such that

$$\alpha = \mathbb{P}_{\lambda_0}[\Lambda \ge Q] \iff \alpha = \mathbb{P}_{\lambda_0}[\tau(X_1, \dots, X_n) \le q]$$

Now, under the null hypothesis, $\tau(X_1, \dots, X_n)$ has a gamma distribution with parameters n and λ_0 (see p. 13). Hence, there exists a q such that $\alpha = \mathbb{P}_{\lambda_0}[\tau(X_1, \dots, X_n) \le q]$, and this q is given by the q_α quantile of the *gamma*(n, λ_0) distribution.

In summary, the optimal test is to reject H_0 at level α if $\tau(X_1, \dots, X_n)$ is smaller than the α-quantile of a *gamma*(n, λ_0) distribution.   □

The previous example demonstrated something interesting: the test statistic for an optimal test reduced to the natural sufficient statistic τ of the distribution (notice that the exponential distribution is a one-parameter exponential family with natural statistic $\tau(x_1, \ldots, x_n) = \sum_{i=1}^{n} x_i$). This is not a coincidence. It works the same way for all one-parameter exponential families:

Example 4.14 (Simple vs Simple Test in Exponential Families)

Let $X_1, \ldots, X_n \overset{iid}{\sim} f(x; \theta)$, where $f(x; \theta) = \exp\{\eta(\theta)T(x) - d(\theta) + S(x)\}$ is a one-parameter exponential family, with η being increasing. Suppose we wish to test $H_0 : \theta = \theta_0$ against $H_1 : \theta = \theta_1$. Without loss of generality, assume that $\theta_0 < \theta_1$. The Neyman–Pearson Lemma (Lemma 4.11, p. 103) dictates that we should look for a test statistic of the form

$$\delta = \mathbf{1}\{L(\theta_1)/L(\theta_0) > Q\} = \mathbf{1}\{\log L(\theta_1) - \log L(\theta_0) > \log Q\}.$$

By the exponential family form of $f(x; \theta)$, we obtain that

$$\delta = \mathbf{1}\left\{(\eta(\theta_1) - \eta(\theta_0)) \sum_{i=1}^{n} T(X_i) - n(d(\theta_1) - d(\theta_0)) > \log Q\right\}$$

$$= \mathbf{1}\left\{\sum_{i=1}^{n} T(X_i) > \frac{\log Q + n(d(\theta_1) - d(\theta_0))}{\eta(\theta_1) - \eta(\theta_0)}\right\}.$$

Notice that $\eta(\theta_1) - \eta(\theta_0) > 0$ since η is increasing, and $n(d(\theta_1) - d(\theta_0))$ is just a constant. So we can just write

$$\delta = \mathbf{1}\{\tau(X_1, \ldots, X_n) > q\},$$

If τ is a continuous random variable, and we want a level α test, then q is going to be the $(1 - \alpha)$-quantile of $G_0(t) = \mathbb{P}_{\theta_0}[\tau(X_1, \ldots, X_n) \le t]$, i.e. the $(1 - \alpha)$-quantile of the sampling distribution of $\tau(X_1, \ldots, X_n)$ when the parameter is taken to be θ_0 (this is called the *null distribution* of τ).

If, on the other hand, we have that η is a decreasing function, then for $\theta_0 < \theta_1$, we have $\eta(\theta_1) - \eta(\theta_0) < 0$. In this case, we can see that the optimal test statistic becomes

$$\delta = \mathbf{1}\{\tau(X_1, \ldots, X_n) \le q\},$$

This time, if we want a level α test, then q must be the α-quantile of $G_0(t) = \mathbb{P}_{\theta_0}[\tau(X_1, \ldots, X_n) \le t]$.

We observe that the form of the test depends on whether η is increasing or decreasing, and on whether $\theta_0 < \theta_1$ or $\theta_0 > \theta_1$. The following table summarises the form of the test statistic for the different cases. In each case, q_s represents the s-quantile of the distribution $G_0(t) = \mathbb{P}_{\theta_0}[\tau(X_1, \ldots, X_n) \le t]$.

	$\theta_0 < \theta_1$	$\theta_0 > \theta_1$
$\eta(\cdot)$ increasing	$\mathbf{1}\{\tau(X_1, \ldots, X_n) > q_{1-\alpha}\}$	$\mathbf{1}\{\tau(X_1, \ldots, X_n) \le q_\alpha\}$
$\eta(\cdot)$ decreasing	$\mathbf{1}\{\tau(X_1, \ldots, X_n) \le q_\alpha\}$	$\mathbf{1}\{\tau(X_1, \ldots, X_n) > q_{1-\alpha}\}$

An interesting observation is that the test function does not depend on the precise value of θ_1, but only on whether or not $\theta_1 < \theta_0$ or $\theta_1 > \theta_0$. $\qquad\square$

It is not always the case that $G_0(t) = \mathbb{P}_{\theta_0}[\tau(X_1, \ldots, X_n) \leq t]$ is a continuous distribution. This means that we might not be able to find an optimal test for all α (we will be able to find an optimal test only for some specific α). Here is an example.

Example 4.15

Let $X_1, \ldots, X_n \overset{iid}{\sim} Poisson(\mu)$ and consider the hypothesis pair

$$H_0 : \mu = \mu_0 \qquad vs \qquad H_1 : \mu = \mu_1.$$

Notice that this is the hypothesis pair we encountered in the Higgs boson example (Example 4.2, p. 96) if we set $\mu_0 = b$ and $\mu_1 = b + s$. This is a one-parameter exponential family example, and it is easy to see that the sufficient statistic is

$$\tau(X_1, \ldots, X_n) = \sum_{i=1}^{n} X_i$$

and the $\eta(\cdot)$ function is strictly increasing (it is equal to the $\log(\cdot)$ function). Since $\mu_1 > \mu_0$, our work in Example 4.14 yields the optimal Neyman–Pearson test statistic for this hypothesis as given by

$$\delta(X_1, \ldots, X_n) = \mathbf{1}\left\{\sum_{i=1}^{n} X_1 > q_{1-\alpha}\right\}.$$

provided there exists a $q_{1-\alpha}$ such that $G_0(q_{1-\alpha}) = \mathbb{P}_{\mu_0}[\tau(X_1, \ldots, X_n) \leq q_{1-\alpha}] = 1 - \alpha$. Since the X_i are independent and Poisson distributed, it is a simple exercise (using generating functions; see Lemma A.10, p. 168) to show that $\tau(X_1, \ldots, X_n) \overset{H_0}{\sim} Poisson(n\mu_0)$. Since this is a discrete distribution, the only α for which this will be the case will be

$$e^{-n\mu_0}, e^{-n\mu_0}(1 + n\mu_0), e^{-n\mu_0}\left(1 + n\mu_0 + \frac{(n\mu_0)^2}{2}\right), e^{-n\mu_0}\left(1 + n\mu_0 + \frac{(n\mu_0)^2}{2} + \frac{(n\mu_0)^3}{3!}\right), \ldots$$

and so on (recall the probability mass function of a Poisson random variable in Definition 1.9, p. 7). However, an interesting observation is that as n grows, this sequence of values becomes denser and denser near to the origin. More precisely, for each $\varepsilon > 0$ and $k \in \mathbb{N}$, there exists an $N \in \mathbb{N}$ such that if $n > N$ then there is at least k possible values of α in the interval $[0, \varepsilon]$. $\square$

Exercise 45 Let $X_1, \ldots, X_n \overset{i.i.d.}{\sim} N(\mu, \sigma^2)$ with $\sigma^2 > 0$ known. Find the most powerful test for the pair $H_0 : \mu = \mu_0$ vs $H_1 : \mu = \mu_1$ with $\mu_0 < \mu_1$ at significance level $\alpha \in (0, 1)$.

Exercise 46 Given a sample $X_1, \ldots, X_n \overset{i.i.d.}{\sim} Bernoulli(p)$, we wish to test

$$H_0 : p = 0.49 \quad vs \quad H_1 : p = 0.51.$$

Determine the (approximate) sample size for which both the probability of type I error and the probability of type II error are equal to 0.01. Use a test function that rejects H_0 when $\sum_i X_i$ is large. Hint: use the central limit theorem, and recall the last part of Exercise 44 (p. 44). You will need to use the fact that $z_{0.99} \approx 2.33$, where $z_{0.99}$ is the 0.99-quantile of the $N(0, 1)$ distribution.

Exercise 47 Let $X_1, \ldots, X_n \overset{\text{i.i.d.}}{\sim} Unif(0, \theta)$ and consider the pair $H_0 : \theta = \theta_0$ and $H_1 : \theta = \theta_1$ with $\theta_1 < \theta_0$.
1. Find the most powerful test of H_0 against H_1 at significance level $\alpha = (\theta_1/\theta_0)^n$. Consider the behaviour of this level as a function of θ_0, θ_1 and n. What is the power of this test? Is it possible to define a Neyman–Pearson optimal test for other values of α?
2. Consider a (not necessarily optimal) test at significance level $\alpha < (\theta_1/\theta_0)^n$ that rejects H_0 when $X_{(n)} < k$. Find the appropriate value of k. What is the power of this test?

Exercise 48 (Intuitive Hypothesis Tests) The goal of this exercise is to motivate hypothesis testing from a more intuitive perspective, via point estimation. Let $X_1, \ldots, X_n$ be iid with density function

$$f_X(x) = \frac{1}{48} \lambda^5 x^{3/2} e^{-\lambda \sqrt{x}}, \qquad x > 0,$$

where $\lambda > 0$ is a parameter. We wish to test $H_0 : \lambda = \lambda_0$ vs. $H_1 : \lambda = \lambda_1$, where $\lambda_0 > \lambda_1$.
1. Find the maximum likelihood estimator $\widehat{\lambda}_n$.
2. As we have seen in Chapter 3, $\widehat{\lambda}_n$ is generally a good estimator. Consequently, a natural approach is to reject H_0 if λ_0 is not "compatible" with $\widehat{\lambda}_n$. In our case this would translate to: reject H_0 if $\widehat{\lambda}_n$ is small. (If it were the case that $\widehat{\lambda}_n > \lambda_0$, we would certainly choose H_0 and not H_1.) What is the form of such a test function (up to a constant, say D)?
3. Now let us find the precise test function. To this aim, we must determine a critical lower bound for $\widehat{\lambda}_n$, sufficiently small to reject H_0. For a given level $\alpha \in (0, 1)$, we wish to choose the lower bound so that the probability of type I error is α. Describe the relationship between the constant D and the level α.
4. We can now wonder whether this is the best test possible. Could we have done better, i.e. find a test at level α but more powerful yet? Show that the answer to this question is in the negative, by proving that our test function is precisely the same as that given by the Neyman–Pearson lemma (you may assume that the value Q in the lemma exists).
5. Find the simplest formula possible for the test function $\delta(X_1, \ldots, X_n)$. Hint: $\widehat{\lambda}_n$ involves a sum, and we know the distribution of each summand.

4.3.2   Unilateral Case

In the case of a unilateral null hypothesis vs a unilateral alternative, there is no result similar to the Neyman–Pearson lemma that describes an optimal test regardless of the specific type of probability model. However, we can still find broad classes of models for which optimal tests can be found. We will not consider the general specifications of such models here, but we note that models that are one-parameter exponential families do satisfy these conditions. Here is the form of the optimal unilateral test in one-parameter exponential families.

Theorem 4.16 (UMP Unilateral Tests for Exponential Families) *Let* $X_1, \ldots, X_n$ *be an iid sample from a one-parameter exponential family with density (or frequency)*

$$f(x; \theta) = \exp\{\eta(\theta)T(x) - d(\theta) + S(x)\}, \qquad x \in \mathcal{X},\ \theta \in \Theta \subseteq \mathbb{R}.$$

where Θ *an open subset, and* $\eta(\cdot)$ *is strictly increasing and continuously differentiable. If* $\tau = \sum_{i=1}^{n} T(X_i)$ *is a continuous random variable, then:*

1. *For* $\alpha \in (0,1)$, *the test statistic* $\delta = \mathbf{1}\{\tau > q_{1-\alpha}\}$ *is uniformly most powerful for testing*

$$\left\{\begin{array}{l} H_0 : \theta \leq \theta_0 \\ H_1 : \theta > \theta_0 \end{array}\right\}$$

   at level α. *Here,* $q_{1-\alpha}$ *is the* $(1-\alpha)$-*quantile of* $G_0(t) = \mathbb{P}_{\theta_0}[\tau \leq t]$.

2. *For* $\alpha \in (0,1)$, *the test statistic* $\delta = \mathbf{1}\{\tau \leq q_\alpha\}$ *is uniformly most powerful for testing*

$$\left\{\begin{array}{l} H_0 : \theta \geq \theta_0 \\ H_1 : \theta < \theta_0 \end{array}\right\}$$

   at level α. *Here,* q_α *is the* α-*quantile of* $G_0(t) = \mathbb{P}_{\theta_0}[\tau \leq t]$.

▶ **Remark 4.17** If $\eta(\cdot)$ is strictly decreasing, then define $\eta_1(\cdot) = -\eta(\cdot)$ and $T_1 = -T$. Then we get an exponential family

$$f(x; \theta) = \exp\{\eta_1(\theta)T_1(x) - d(\theta) + S(x)\}, \qquad x \in \mathcal{X},\ \theta \in \Theta \subseteq \mathbb{R},$$

with $\eta_1(\cdot)$ strictly increasing. The theorem now applies in the same way, using $\tau_1 = \sum_{i=1}^{n} T_1(X_i)$ in lieu of τ. We can summarise the form of the test statistic, as dependent on the direction of the hypotheses and on whether η is increasing or decreasing in the following table:

	$\begin{pmatrix} H_0 : \theta \le \theta_0 \\ H_1 : \theta > \theta_0 \end{pmatrix}$	$\begin{pmatrix} H_0 : \theta \ge \theta_0 \\ H_1 : \theta < \theta_0 \end{pmatrix}$
$\eta(\cdot)$ increasing	$\mathbf{1}\{\tau(X_1,\ldots,X_n) > q_{1-\alpha}\}$	$\mathbf{1}\{\tau(X_1,\ldots,X_n) \le q_\alpha\}$
$\eta(\cdot)$ decreasing	$\mathbf{1}\{\tau(X_1,\ldots,X_n) \le q_\alpha\}$	$\mathbf{1}\{\tau(X_1,\ldots,X_n) > q_{1-\alpha}\}$

▶ **Remark 4.18** Notice that, surprisingly, the form of the test is exactly the same as the form of the test in an exponential family for a "simple vs simple" hypothesis pair (compare the table above with the table in Example 4.14, p. 106). How is this possible? The key observation is that, as we saw in Example 4.14, the form of the Neyman–Pearson test function did not depend on the precise value of θ_1, but only on whether or not $\theta_1 < \theta_0$ or $\theta_1 > \theta_0$. It also depended on θ_0. This explains why the form of the test function in the unilateral case is the same as in the simple vs simple case. This is not the true in general, but is true for one-parameter exponential families due to their special form.

Proof of Theorem 4.16 We will prove part (1), since part (2) follows directly analogously. To prove (1), we need to verify two things:

(I) That $\sup_{\theta \in (-\infty,\theta_0]} \mathbb{P}_\theta[\delta = 1] \le \alpha$ (i.e. that δ maintains level α over the entire null parameter space). Note that since δ is a Bernoulli random variable, $\mathbb{P}_\theta[\delta = 1] = \mathbb{E}_\theta[\delta]$.

(II) That for any $\psi : \mathcal{X}^n \to \{0,1\}$ such that $\sup_{\theta \in (-\infty,\theta_0]} \mathbb{P}_\theta[\psi = 1] \le \alpha$, it must be that

$$\mathbb{E}_\theta[\psi] \le \mathbb{E}_\theta[\delta], \quad \forall \theta \in (\theta_0, \infty).$$

(i.e. that δ has maximal power over the entire alternative parameter space).

The key to showing (I) is to show that $\theta \mapsto \mathbb{E}_\theta[\delta(X_1,..,X_n)] = \mathbb{P}_\theta[\delta = 1]$ is increasing, by showing that its derivative is non-negative. Since $\eta(\cdot)$ and $d(\cdot)$ are differentiable, $f(x;\theta)$ is of the exponential family form and $\delta : \mathcal{X} \to \{0,1\}$, we may differentiate under the integral (see Remark 3.11, p. 67),

$$\frac{\partial}{\partial\theta}\mathbb{E}_\theta[\delta] = \frac{\partial}{\partial\theta} \int_{\mathcal{X}^n} \delta(x_1,\ldots,x_n) \prod_{i=1}^n f(x_i;\theta)dx_1\ldots dx_n$$

$$= \int_{\mathcal{X}^n} \delta(x_1,\ldots,x_n)\frac{\partial}{\partial\theta} \prod_{i=1}^n f(x_i;\theta)dx_1\ldots dx_n$$

$$= \int_{\mathcal{X}^n} \delta(x_1,\ldots,x_n) \left(\frac{\prod_{i=1}^n f(x_i;\theta)}{\prod_{i=1}^n f(x_i;\theta)}\right) \frac{\partial}{\partial\theta} \prod_{i=1}^n f(x_i;\theta)dx_1\ldots dx_n$$

$$= \int_{\mathcal{X}^n} \delta(x_1, \ldots, x_n) \left(\frac{\partial}{\partial \theta} \log \prod_{i=1}^n f(x_i; \theta) \right) \prod_{i=1}^n f(x_i; \theta) dx_1 \ldots dx_n$$

$$= \mathbb{E}_\theta \left[\delta(X_1, \ldots, X_n) \sum_{i=1}^n \frac{\partial}{\partial \theta} \log f(X_i; \theta) \right]$$

$$= \mathrm{Cov}_\theta \left[\delta(X_1, \ldots, X_n), \sum_{i=1}^n \frac{\partial}{\partial \theta} \log f(X_i; \theta) \right]$$

$$= \mathrm{Cov}_\theta \left[\delta(X_1, \ldots, X_n), \left(\eta'(\theta)\tau(X_1, \ldots, X_n) - nd'(\theta) \right) \right]$$

$$= \eta'(\theta)\mathrm{Cov}_\theta[\delta, \tau]$$

The third to last equality comes from the fact that when we can differentiate under the integral sign,[1]

$$\mathbb{E}_\theta \left[\sum_{i=1}^n \frac{\partial}{\partial \theta} \log f(X_i; \theta) \right] = 0.$$

In the discrete case, we simply replace integration by summation, of course. With this result under our belts, we can now verify (I). Notice first that $\mathbb{P}_{\theta_0}[\delta = 1] = \mathbb{P}_{\theta_0}[\tau > q_{1-\alpha}] = 1 - \mathbb{P}_{\theta_0}[\tau \leq q_{1-\alpha}]$. But $1 - \mathbb{P}_{\theta_0}[\tau \leq q_{1-\alpha}] = 1 - G_0(q_{1-\alpha}) = 1 - (1 - \alpha) = \alpha$, since $q_{1-\alpha}$ is the $(1 - \alpha)$-quantile of G_0. Further to this, we have calculated that $\frac{\partial}{\partial \theta}\mathbb{P}_\theta[\delta = 1] = \frac{\partial}{\partial \theta}\mathbb{E}_\theta[\delta] = \eta'(\theta)\mathrm{Cov}_\theta[\delta, \tau]$. But $\eta'(\theta)$ is positive since $\eta(\cdot)$ is increasing, and $\mathrm{Cov}_\theta[\delta, \tau] \geq 0$ because $\delta = \mathbf{1}\{\tau > q_{1-\alpha}\}$ is an increasing function of τ, and is thus positively correlated with τ (see Lemma A.5, p. 160)). It follows that $\frac{\partial}{\partial \theta}\mathbb{P}_\theta[\delta = 1] \geq 0$, and so $\mathbb{P}_\theta[\delta = 1]$ is increasing. It must thus be that $\mathbb{P}_\theta[\delta = 1] \leq \mathbb{P}_{\theta_0}[\delta = 1] = \alpha$ for all $\theta < \theta_0$, and the proof of part (I) is complete.

To prove part (II), let θ_1 be an arbitrary element in (θ_0, ∞). Notice that

$$\Lambda := \frac{f(X_1, \ldots, X_n; \theta_1)}{f(X_1, \ldots, X_n; \theta_0)} = \exp\{\eta(\theta_1)\tau - nd(\theta_1) - \eta(\theta_0)\tau + nd(\theta_0)\}$$

$$= \exp\{[\eta(\theta_1) - \eta(\theta_0)]\tau - nd(\theta_1) + nd(\theta_0)\}$$

It follows that the likelihood ratio is a strictly monotone function of τ, since $\eta(\cdot)$ is strictly increasing. Therefore, δ is equal to the likelihood ratio test function

$$\mathbf{1}\Big\{\Lambda > \underbrace{\exp\{[\eta(\theta_1) - \eta(\theta_0)]q_{1-\alpha} - nd(\theta_1) + nd(\theta_0)\}}_{\varrho}\Big\}$$

---

[1]To verify this, replace δ by 1 in the array of equations right above.

since δ is 1 if and only if $\mathbb{1}\{\Lambda > Q\}$ is 1. It follows from the Neyman–Pearson lemma (Lemma 4.11, p. 103) that

$$\mathbb{E}_{\theta_1}[\psi] \leq \mathbb{E}_{\theta_1}[\delta], \quad \forall \theta_1 \in (\theta_0, \infty).$$

for any $\psi : \mathcal{X}^n \to \{0, 1\}$ such that $\mathbb{P}_{\theta_0}[\psi = 1] \leq \alpha$. Note, however, that

$$\sup_{\theta \leq \theta_0} \mathbb{P}_\theta[\psi = 1] \leq \alpha \implies \mathbb{P}_{\theta_0}[\psi = 1] \leq \alpha.$$

And so, from what we have just proven, $\sup_{\theta \leq \theta_0} \mathbb{P}_\theta[\psi = 1] \leq \alpha$ must thus imply

$$\mathbb{E}_{\theta_1}[\psi] \leq \mathbb{E}_{\theta_1}[\delta], \quad \forall \theta_1 \in (\theta_0, \infty).$$

This proves (II) and thus the proof is complete.   □

Exercise 49 A bio-imaging laboratory has developed a new method to carry out brain scans in less than 20 min. A sample of 12 brain scan durations from the lab is given below:

$$\mathbb{X} = \{21, 18, 19, 16, 18, 24, 22, 19, 24, 26, 18, 21\}.$$

1. Suppose that the duration time approximately follows an $N(\mu, 3^2)$ distribution. Test whether the mean scan time is less than 20 min, i.e., test $H_0 : \mu \leq \mu_0$ vs $H_1 : \mu > \mu_0$ with $\mu_0 = 20$ at significance level $\alpha = 0.05$.
2. Could you carry out the same analysis if the variance were unknown?
   Hint:use $\delta = \mathbb{1}\left(\frac{\sqrt{n}(\bar{X} - \mu_0)}{S} \geq t_{n-1,1-\alpha}\right)$ as your test function. Here $t_{n-1,1-\alpha}$ is the $1 - \alpha$ quantile of the Student t distribution with $n - 1$ degrees of freedom.

Exercise 50 Let $Y_1, \ldots, Y_4$ be iid $N(\mu, 4^2)$ random variables. We wish to determine whether μ is larger than $\mu_0 = 10$. To this aim, we carry out a test at level $\alpha = 5\%$ contrasting the hypotheses $H_0 : \mu \leq 10$ and $H_1 : \mu > 10$.
1. Calculate the power of the test when the true value of μ equals 13 and when it equals 11.
2. Determine what number of observations we need to have to guarantee that the power of the test is at least 90% when the true mean is $\mu = 13$.

Exercise 51 (Paired Test) A standard problem in the pharmaceutical industry is to determine whether treatment with a new drug will have an effect on a patient. Consider, for instance, the problem of reducing blood pressure, perhaps even by placebo effect. Let X_i be the blood pressure of the ith patient before the drug treatment, and Y_i the ith patient's blood pressure at the end of the treatment. We may suppose that the X_i are iid, since different patients are chosen at random. Similarly, the Y_i are independent, since all patients received the same treatment.

Assume that $X_i \sim N(\mu_1, \sigma_1^2)$ and $Y_i \sim N(\mu_2, \sigma_2^2)$, with unknown σ_1^2, σ_2^2. Construct a test in order to test the hypothesis that the drug treatment lowers the blood pressure. Remark: since X_i and Y_i come from the same person (patient i), we cannot assume them to be independent. In this context, we speak of a paired test.

4.3.2.1 Approximate Critical Values

Notice that, in order to be able to implement the unilateral test in practice, we will need to know how to calculate the quantile q_s in the table above. This can be calculated provided that $G_0(t) = \mathbb{P}_{\theta_0}[\tau(X_1, \ldots, X_n) \leq t]$ is known exactly. In the examples we considered (e.g. Example 4.13) this was indeed the case, but it will not always be the case: as we saw in Sect. 2.4 (p. 53) it is often not possible to determine the precise distribution $G_0(t)$. However, one can approximate it for large values of the sample size n. Specifically, Corollary 2.24 (p. 56) tells us that

$$\sqrt{n}(n^{-1}\tau(X_1, \ldots, X_n) - \gamma'(\phi)) \overset{d}{\longrightarrow} N(0, \gamma''(\phi)),$$

or, equivalently, by Exercise 23 (p. 52)

$$\sqrt{n}\left(n^{-1}\tau(X_1, \ldots, X_n) - \frac{d'(\theta)}{\eta'(\theta)}\right) \overset{d}{\longrightarrow} N\left(0, \frac{d''(\theta)\eta'(\theta) - d'(\theta)\eta''(\theta)}{[\eta'(\theta)]^3}\right).$$

The latter suggests approximating the distribution $G_0(t) = \mathbb{P}_{\theta_0}[\tau(X_1, \ldots, X_n) \leq t]$ by a

$$N\left(n\frac{d'(\theta_0)}{\eta'(\theta_0)}, \, n\frac{d''(\theta_0)\eta'(\theta_0) - d'(\theta_0)\eta''(\theta_0)}{[\eta'(\theta_0)]^3}\right)$$

distribution, when n is sufficiently large. Notice that since this latter distribution is a continuous distribution, it follows that for large enough samples from exponential families, we are able to approximately construct the Neyman–Pearson optimal test for any level α. This can be done by the method of standardisation (Lemma 1.32, p. 22), thus employing tables of quantiles for the $N(0, 1)$ distribution.

4.3.3   Bilateral Case

Unfortunately, for hypothesis pairs of the form $H_0 : \theta = \theta_0$ and $H_1 : \theta \neq \theta_0$, there can be no optimal test, in the sense described in Definition 4.10 (p. 103). To see this, note that for $\delta : \mathcal{X}^n \to \{0, 1\}$ to be uniformly most powerful for $H_0 : \theta = \theta_0$ vs $H_1 : \theta \neq \theta_0$, it must be most powerful for $H_0 : \theta = \theta_0$ and $H_1 : \theta = \theta_1$, for all $\theta_1 \neq \theta_0$. But consider the problem of testing such a pair, in a one-parameter exponential family $f(x; \theta) = \exp\{\eta(\theta)T(x) - d(\theta) + S(x)\}$. Example 4.14 (p. 106) tells us that the form of the test is different depending on whether $\theta_1 > \theta_0$ or $\theta_1 < \theta_0$, so there can be no optimal test: if a test is most powerful over (θ_0, ∞), then it will necessarily be less powerful than some other test over $(-\infty, \theta_0)$.

Because of this, we need to abandon the hope of uniquely determining the best testing method, as we were able to do in the previous two paragraphs. Instead, we must look for testing methods that will yield reasonably well-performing tests in general. We will consider two such methods, both motivated by the notion of likelihood: The *Likelihood Ratio Method* and *Wald's Method*.

4.3.3.1 Likelihood Ratio Tests

In the previous chapter we saw that the concept of likelihood is of fundamental importance in the problem of point estimation. In particular, we saw that we can construct estimators with excellent properties if we use the method of maximum likelihood: choosing as our estimator the element of the parameter space that maximises the likelihood.

The motivation behind the likelihood ratio test is to use the concept of likelihood again, but this time in order to decide between the two competing hypotheses. The hope is that such an approach will yield powerful tests. The formal definition is as follows.

Definition 4.19 (Likelihood Ratio Test)

Let $X_1, \ldots, X_n \overset{iid}{\sim} f(x; \theta)$, yielding a likelihood

$$L(\theta) = \prod_{i=1}^{n} f(X_i; \theta),$$

and let $H_0 : \theta \in \Theta_0$ and $H_1 : \theta \in \Theta_1$ be two competing hypotheses. Define the likelihood ratio as

$$\Lambda(X_1, \ldots, X_n) = \frac{\sup_{\theta \in \Theta_1} L(\theta)}{\sup_{\theta \in \Theta_0} L(\theta)}.$$

The Likelihood Ratio Test (LRT) at level $\alpha \in (0, 1)$ is defined to be the test with test function

$$\delta(X_1, \ldots, X_n) = \mathbf{1}\{\Lambda(X_1, \ldots, X_n) > Q\},$$

where $Q > 0$ is such that $\sup_{\theta \in \Theta_0} \mathbb{P}_\theta[\Lambda(X_1, \ldots, X_n) > Q] = \alpha$, provided such a Q exists.

What is the intuition behind the LRT? When we had a simple vs simple hypothesis pair, the Neyman–Pearson Lemma (Lemma 4.11, p. 103) said that we should compare the likelihood evaluated at the alternative value θ_1 to the likelihood evaluated at the null value θ_0. When either of these sets may not be a singleton, the LRT method suggests that we simply compare the maximum achievable likelihood from within Θ_1 to the maximum achievable likelihood from within Θ_0, thus mimicking the Neyman–Pearson lemma.

► **Remark 4.20 (LRT for Bilateral Hypothesis Pairs)**   Note that when $H_0 : \theta = \theta_0$ and $H_1 : \theta \neq \theta_0$ we have $\Theta_0 = \{\theta_0\}$ and $\Theta_1 = \mathbb{R} \setminus \{\theta_0\}$, so, if L is a continuous function of θ and attains its supremum,

$$\Lambda(X_1, \ldots, X_n) = \frac{\sup_{\theta \in \Theta_1} L(\theta)}{\sup_{\theta \in \Theta_0} L(\theta)} = \frac{\sup_{\theta \in \mathbb{R} \setminus \{\theta\}} L(\theta)}{L(\theta_0)} = \frac{\sup_{\theta \in \mathbb{R}} L(\theta)}{L(\theta_0)} = \frac{L(\hat{\theta})}{L(\theta_0)},$$

where $\hat{\theta}$ is a maximum likelihood estimator of θ.

Example 4.21

Let $X_1, \ldots, X_n \overset{iid}{\sim} N(\mu, \sigma^2)$. Assume that σ^2 is known and suppose we are interested in testing the hypothesis pair

$$H_0 : \mu = \mu_0 \quad \text{vs} \quad H_1 : \mu \neq \mu_0.$$

Since the MLE of μ is $\bar{X}$, we have

$$L(\bar{X}) = \left(\frac{1}{2\pi\sigma^2} \right)^{n/2} \exp\left\{ -\frac{1}{2\sigma^2} \sum_{i=1}^{n} (X_i - \bar{X})^2 \right\}, \quad \&$$

$$L(\mu_0) = \left(\frac{1}{2\pi\sigma^2} \right)^{n/2} \exp\left\{ -\frac{1}{2\sigma^2} \sum_{i=1}^{n} (X_i - \mu_0)^2 \right\}.$$

Consequently,

$$\Lambda(X_1, \ldots, X_n) = \frac{L(\hat{\mu})}{L(\mu_0)} = \frac{L(\bar{X})}{L(\mu_0)} = \exp\left\{ -\frac{1}{2\sigma^2} \left[\sum_{i=1}^{n} (X_i - \bar{X})^2 - \sum_{i=1}^{n} (X_i - \mu_0)^2 \right] \right\}.$$

But we note that

$$\sum_{i=1}^{n} (X_i - \mu_0)^2 = \sum_{i=1}^{n} (X_i - \bar{X} + \bar{X} - \mu_0)^2 = \sum_{i=1}^{n} (X_i - \bar{X})^2 + n(\bar{X} - \mu_0)^2,$$

because the cross-terms vanish. It follows that the likelihood ratio reduces to

$$\Lambda(X_1, \ldots, X_n) = \exp\left\{ \frac{n}{2\sigma^2} (\bar{X} - \mu_0)^2 \right\}.$$

It follows that $\Lambda(X_1, \ldots, X_n)$ is a monotone increasing function of $S(X_1, \ldots, X_n) = \left(\frac{\bar{X} - \mu_0}{\sigma/\sqrt{n}} \right)^2$. Note that when H_0 is true, $S \sim \chi_1^2$ (recall Example 1.29, p. 20). Therefore, the likelihood ratio test rejects the null hypothesis if and only if $S(X_1, \ldots, X_n) > \chi_{1,1-\alpha}^2$, where $\chi_{1,1-\alpha}^2$ denotes the

$1 - \alpha$ quantile of the χ_1^2 distribution. Notice that this is equivalent to rejecting the null if and only if $\left|\frac{\bar{X} - \mu_0}{\sigma/\sqrt{n}}\right| > z_{1-\alpha/2}$, where $z_{1-\alpha/2}$ is the $(1 - \alpha/2)$ quantile of an $N(0, 1)$ distribution.   □

An important aspect of the Likelihood Ratio method is that it can handle situations where there are more than one parameters, but we are interested in testing a bilateral hypothesis for a single parameter. In other words, suppose that $X_1, \ldots, X_n \overset{iid}{\sim} f(x; \theta, \xi)$, where $\theta \in \mathbb{R}$ and $\xi \in \mathbb{R}^p$ are two unknown parameters. We might be interested in testing

$$H_0 : \theta = \theta_0 \qquad \text{vs} \qquad H_1 : \theta \neq \theta_0$$

at level $\alpha > 0$, for some $\theta_0 \in \mathbb{R}$, without making any reference to (and without caring about) the remaining parameter ξ (a parameter such as ξ is often referred to as a *nuisance parameter*). In this case, the likelihood ratio is formed as

$$\Lambda(X_1, \ldots, X_n) = \frac{\sup_{\theta \in \mathbb{R} \setminus \{\theta_0\}, \xi \in \mathbb{R}^p} L(\theta, \xi)}{\sup_{\theta \in \{\theta_0\}, \xi \in \mathbb{R}^p} L(\theta, \xi)} = \frac{\sup_{\theta \in \mathbb{R}, \xi \in \mathbb{R}^p} L(\theta, \xi)}{\sup_{\xi \in \mathbb{R}^p} L(\theta_0, \xi)} = \frac{L(\hat{\theta}, \hat{\xi})}{\sup_{\xi \in \mathbb{R}^p} L(\theta_0, \xi)},$$

where $(\hat{\theta}, \hat{\xi})$ is an MLE of (θ, ξ). The Likelihood Ratio Test at level $\alpha \in (0, 1)$ will be defined again as the test with test function

$$\delta(X_1, \ldots, X_n) = \mathbf{1}\{\Lambda(X_1, \ldots, X_n) > Q\},$$

where $Q > 0$ is such that $\sup_{\xi \in \mathbb{R}^p} \mathbb{P}_{\theta_0, \xi}[\Lambda(X_1, \ldots, X_n) > Q] = \alpha$, provided such a Q exists. Here is the classical example:

Example 4.22 (Bilateral Test for Means of Gaussian Distributions)

Let $X_1, \ldots, X_n \overset{iid}{\sim} N(\mu, \sigma^2)$, where μ and σ^2 are unknown. Suppose we wish to test the hypothesis pair

$$H_0 : \mu = \mu_0 \qquad \text{vs} \qquad H_1 : \mu \neq \mu_0$$

at level $\alpha > 0$, for some fixed $\mu_0 \in \mathbb{R}$. Let us use the Likelihood Ratio method in order to derive a suitable test. We notice that we have two parameters, but are only interested in one of them. Following the reasoning presented above, we need to determine

$$\Lambda(X_1, \ldots, X_n) = \frac{L(\hat{\mu}, \hat{\sigma}^2)}{\sup_{\sigma^2 > 0} L(\mu_0, \sigma^2)}, \tag{4.2}$$

where $(\hat{\mu}, \hat{\sigma}^2)$ is the MLE of (μ, σ^2). For the numerator, one may calculate that

$$\frac{\partial}{\partial \sigma^2} \ell(\mu_0, \sigma^2) = -\frac{n}{2\sigma^2} + \frac{1}{2\sigma^4} \sum_{i=1}^{n} (X_i - \mu_0)^2.$$

Following the same steps as in Exercise 3.16 (p. 71), we conclude that

$$\arg \sup_{\sigma^2 > 0} L(\mu_0, \sigma^2) = \frac{1}{n} \sum_{i=1}^{n} (X_i - \mu_0)^2.$$

In other words, the supremum in the numerator in Eq. (4.2) satisfies

$$\sup_{\sigma^2 > 0} L(\mu_0, \sigma^2) = L\left(\mu_0, \frac{1}{n} \sum_{i=1}^{n} (X_i - \mu_0)^2\right),$$

and so the numerator is equal to

$$\sup_{\sigma^2 > 0} L(\mu_0, \sigma^2) = \left[\frac{1}{2\pi(1/n) \sum_{i=1}^{n}(X_i - \mu_0)^2}\right]^{n/2} \exp\left\{-\frac{\sum_{i=1}^{n}(X_i - \mu_0)^2}{(2/n) \sum_{i=1}^{n}(X_i - \mu_0)^2}\right\}$$

$$= \left[\frac{ne^{-1}}{2\pi \sum_{i=1}^{n}(X_i - \mu_0)^2}\right]^{n/2}.$$

Next, we turn to the denominator in Eq. (4.2). Recalling Example 3.16 (p. 71), we have that the MLE of (μ, σ^2) is given by the pair:

$$\hat{\mu} = \frac{1}{n} \sum_{i=1}^{n} X_i = \overline{X}, \qquad \& \qquad \hat{\sigma}^2 = \frac{1}{n} \sum_{i=1}^{n} (X_i - \overline{X})^2.$$

It follows that

$$L(\hat{\mu}, \hat{\sigma}^2) = \left[\frac{1}{2\pi(1/n) \sum_{i=1}^{n}(X_i - \overline{X})^2}\right]^{n/2} \exp\left\{-\frac{\sum_{i=1}^{n}(X_i - \overline{X})^2}{(2/n) \sum_{i=1}^{n}(X_i - \overline{X})^2}\right\}$$

$$= \left[\frac{ne^{-1}}{2\pi \sum_{i=1}^{n}(X_i - \overline{X})^2}\right]^{n/2}.$$

Consequently, the likelihood ratio is

$$\Lambda(X_1, \ldots, X_n) = \frac{L(\hat{\mu}, \hat{\sigma}^2)}{\sup_{\sigma^2 > 0} L(\mu_0, \sigma^2)} = \left[\frac{\sum_{i=1}^{n}(X_i - \mu_0)^2}{\sum_{i=1}^{n}(X_i - \overline{X})^2}\right]^{n/2}.$$

This can be further simplified by recalling that

$$\sum_{i=1}^{n}(X_i - \mu_0)^2 = \sum_{i=1}^{n}(X_i - \overline{X} + \overline{X} - \mu_0)^2 = \sum_{i=1}^{n}(X_i - \overline{X})^2 + n(\overline{X} - \mu_0)^2,$$

since the cross-terms vanish. Using this fact, we may write

$$\Lambda(X_1, \ldots, X_n) = \left[\frac{\sum_{i=1}^{n}(X_i - \overline{X})^2 + n(\overline{X} - \mu_0)^2}{\sum_{i=1}^{n}(X_i - \overline{X})^2}\right]^{n/2} = \left\{1 + \frac{n(\overline{X} - \mu_0)^2}{\sum_{i=1}^{n}(X_i - \overline{X})^2}\right\}^{n/2}.$$

Observe now that

$$\Lambda > Q \iff \underbrace{\frac{n(\overline{X} - \mu_0)^2}{\sum_{i=1}^n (X_i - \overline{X})^2/(n-1)}}_{T^2} > \underbrace{(n-1)(Q^{2/n} - 1)}_{:=C} \iff \underbrace{\left| \frac{\overline{X} - \mu_0}{S/\sqrt{n}} \right|}_{|T|} > \sqrt{C},$$

so the likelihood ratio test is

$$\delta(X_1, \ldots, X_n) = \mathbf{1}\{\Lambda > Q\} = \mathbf{1}\left\{ \left| \frac{\overline{X} - \mu_0}{S/\sqrt{n}} \right| > \sqrt{C} \right\},$$

and $\sqrt{C}$ needs to be selected so that $\mathbb{P}_{H_0}\left[\left| \frac{\overline{X} - \mu_0}{S/\sqrt{n}} \right| > \sqrt{C} \right] = \alpha$. But, when H_0 is true, we have that $T \sim t_{n-1}$, the latter denoting Student's distribution with $n-1$ degrees of freedom (see Theorem 2.9, p. 48). It follows that $\sqrt{C} = t_{n-1,1-\alpha/2}$, where the latter is the $(1 - \alpha/2)$-quantile of the t_{n-1} distribution. In conclusion, the LRT is

$$\delta = \mathbf{1}\left\{ |\overline{X} - \mu_0| > t_{n-1,1-\alpha/2} S/\sqrt{n} \right\}.$$

$\square$

Notice the intuition in this result: we will reject the hypothesis $H_0 : \mu = \mu_0$ if $\overline{X}$ (the MLE of μ) is at a "significant" distance from μ_0. How large is "significant"? The answer is $t_{n-1,1-\alpha/2}$ times the (estimated) standard deviation of $\overline{X}$ (estimated by $S/\sqrt{n}$). We will see in Sect. 4.3.3.3 that we can motivate another type of testing method by generalising this idea. For the moment, though, we turn to consider another important problem in the next section.

Exercise 52 (Bilateral Test for Variances of Gaussian Distributions) Let $X_1, \ldots, X_n$ be an iid random sample from a $\mathcal{N}(\mu, \sigma^2)$ distribution, where both μ and σ^2 are unknown. Show that the LRT for the hypothesis pair $H_0 : \sigma^2 = \sigma_0^2$ vs $H_1 : \sigma^2 \neq \sigma_0^2$ Ã at level a α is of the form $\mathbf{1}\{W > c_1\} + \mathbf{1}\{W < c_2\}$, where $W = (1/\sigma_0^2) \sum_{i=1}^n (X_i - \overline{X})^2$, and c_1 and c_2 are such that $c_1^{-n} e^{c_1} = c_2^{-n} e^{c_2}$.
    Hint : Write the likelihood ratio as a function of W and investigate the form of this function. Remark: in practice, one usually chooses c_1 and c_2 such that $\mathbb{P}_{H_0}(W > c_1) = \mathbb{P}_{H_0}(W < c_2) = \alpha/2$ (which is no longer a likelihood ratio test.)

Exercise 53 (Unpaired Test) Let $X_1, \ldots, X_n, Y_1, \ldots, Y_m$ be a sample of $n + m$ independent random variables, where $X_i \overset{iid}{\sim} N(\mu_1, \sigma^2)$ and $Y_i \overset{iid}{\sim} N(\mu_2, \sigma^2)$, and σ^2 is unknown (but the same for the X and the Y). The goal of this exercise is to determine the LRT for the hypothesis pair $H_0 : \mu_1 = \mu_2$ vs $H_1 : \mu_1 \neq \mu_2$.
1. Define the likelihood of the parameter $\theta = (\mu_1, \mu_2, \sigma^2)$.
2. Noting that $\Theta_0 = \{(\mu, \mu, \sigma^2) : -\infty < \mu < \infty, 0 < \sigma^2 < \infty\}$ and $\Theta_1 = \{(\mu_1, \mu_2, \sigma^2) : -\infty < \mu_1 \neq \mu_2 < \infty, 0 < \sigma^2 < \infty\}$, show that

$$\sup_{\theta \in \Theta_0} L(\theta) = \left(\frac{e^{-1}}{2\pi \hat{\sigma}_{\Theta_0}^2} \right)^{(m+n)/2},$$

where $\hat{\sigma}^2_{\Theta_0} = \frac{1}{n+m}\left(\sum_{i=1}^{n}(X_i - \hat{\mu})^2 + \sum_{j=1}^{m}(Y_j - \hat{\mu})^2\right)$, with $\hat{\mu} = \frac{1}{n+m}\left(\sum_{i=1}^{n}X_i + \sum_{j=1}^{m}Y_j\right)$.

Show further that

$$\sup_{\theta \in \Theta_1} L(\theta) = \left(\frac{e^{-1}}{2\pi\hat{\sigma}^2_{\Theta_1}}\right)^{(m+n)/2},$$

where $\hat{\sigma}^2_{\Theta_1} = \frac{1}{n+m}\left(\sum_{i=1}^{n}(X_i - \bar{X})^2 + \sum_{j=1}^{m}(Y_j - \bar{Y})^2\right)$.

3. Using the fact that $\sum_{i=1}^{n}(X_i - \hat{\mu})^2 = \sum_{i=1}^{n}(X_i - \bar{X})^2 + \frac{nm^2(\bar{X}-\bar{Y})^2}{(n+m)^2}$ and that $\sum_{j=1}^{m}(Y_j - \hat{\mu})^2 = \sum_{j=1}^{m}(Y_j - \bar{Y})^2 + \frac{mn^2(\bar{X}-\bar{Y})^2}{(n+m)^2}$, show that

$$\Lambda(X_1,\ldots,X_n,Y_1,\ldots,Y_m) = \left(1 + \frac{T^2}{m+n-2}\right)^{(n+m)/2},$$

where

$$T = \frac{\sqrt{\frac{nm}{n+m}}(\bar{X} - \bar{Y})}{\sqrt{\frac{1}{n+m-2}[(n-1)S_X^2 + (m-1)S_Y^2]}},$$

with $S_X^2 = \frac{1}{n-1}\sum_{i=1}^{n}(X_i - \bar{X})^2$ and $S_Y^2 = \frac{1}{m-1}\sum_{j=1}^{m}(Y_j - \bar{Y})^2$.

4. Using the fact that the level α test with test function given by $\mathbf{1}\{\Lambda(X_1,\ldots,X_n, Y_1,\ldots,Y_m) > Q\}$ is the same as the level α test with test function $\mathbf{1}\{|T| > Q'\}$ where Q' is such that $\sup_{\theta \in \Theta_0} \mathbb{P}_\theta(|t| > Q') = \alpha$, determine the LRT, i.e. find the law of T under H_0 as well as the value of Q'.

Hint: if $A \sim \chi_a^2$ and $B \sim \chi_b^2$ are independent, it follows that $A + B \sim \chi_{a+b}^2$. Theorem 2.9 (p. 48) could also be useful.

4.3.3.2 Approximate Critical Values for Likelihood Ratio Tests

In Example (4.22) we were able to find the precise value of Q needed in the LRT statistic $\delta = \mathbf{1}\{\Lambda > Q\}$, by reducing the test statistic to an equivalent expression, and by using the properties of the normal distribution. This may not be the case more generally, though, where we may not be able to find the exact distribution of Λ (or a monotone function of it) and thus derive the exact Q. In these cases, we will need to resort to large sample approximations, as we have done in other cases where exact sampling distributions were not available. We consider the problem of finding the approximate distribution of Λ under simple nulls for one-parameter exponential families.

Theorem 4.23 *Let $X_1, \ldots, X_n$ be an iid sample from a distribution with density (or mass function) $f(x; \theta)$ which belongs to a non-degenerate one-parameter exponential family,*

$$f(x; \theta) = \exp\{\eta(\theta)T(x) - d(\theta) + S(x)\}, \qquad x \in \mathcal{X}, \, \theta \in \Theta$$

Assume that:
1. *The parameter space $\Theta \subset \mathbb{R}$ is an open set.*
2. *The function $\eta(\cdot)$ is a twice continuously differentiable bijection between Θ and $\Phi = \eta(\Theta)$.*

Let $\hat{\theta}_n$ be the maximum likelihood estimator of θ, and $\theta_0 \in \Theta$ be some fixed element of the parameter space such that $\eta'(\theta_0) \neq 0$. If $\Lambda(X_1, \ldots, X_n) = L(\hat{\theta})/L(\theta_0)$ is the likelihood ratio, then

$$2 \log \Lambda(X_1, \ldots, X_n) = 2(\ell(\hat{\theta}) - \ell(\theta_0)) \xrightarrow{d} \chi_1^2,$$

whenever $\{H_0 : \theta = \theta_0\}$ is true.

▶ **Remark 4.24 (Likelihood Ratio vs LogLikelihood Difference)** Notice that knowing the distribution of $2 \log \Lambda$ under the null hypothesis is equivalent to knowing the distribution of Λ under the null hypothesis, since the mapping $x \mapsto 2 \log x$ is monotone. The result above can thus be used in order to determine the right critical value for a likelihood ratio test. Specifically, the likelihood ratio test function $\mathbf{1}\{\Lambda > Q\}$ will be approximately (for n being large) equivalent to the test function

$$\mathbf{1}\{2 \log \Lambda > \chi_{1,1-\alpha}^2\} = \mathbf{1}\left\{\Lambda > \exp\left(\frac{\chi_{1,1-\alpha}^2}{2}\right)\right\},$$

where $\chi_{1,1-\alpha}^2$ denotes the $(1 - \alpha)$-quantile of the χ_1^2 distribution. In other words, for large n, the approximate critical value should be $Q \approx \exp\left(\frac{\chi_{1,1-\alpha}^2}{2}\right)$.

Proof of Theorem 4.23 We apply a second order Taylor expansion with Lagrange form of remainder (Theorem A.1, p. 159) to obtain

$$2(\ell(\hat{\theta}_n) - \ell(\theta_0)) = 2\ell'(\hat{\theta}_n)(\hat{\theta}_n - \theta_0) - \ell''(\theta_n^*)(\hat{\theta}_n - \theta_0)^2 = [d''(\theta_n^*) - \eta''(\theta_n^*)\overline{T}][\sqrt{n}(\hat{\theta}_n - \theta_0)]^2,$$

where θ_n^* is between $\hat{\theta}_n$ and θ_0, and $\ell'(\hat{\theta}_n) = 0$ since $\hat{\theta}$ maximises the likelihood. It follows that $|\theta_n^* - \theta_0| \leq |\hat{\theta}_n - \theta_0|$, and thus $\theta_n^* \xrightarrow{p} \theta_0$ by consistency of $\hat{\theta}_n$. We now consider the behaviour of the terms involved in the Taylor expansion as $n \to \infty$. The continuous mapping theorem (Theorem 2.25 p. 57) implies that $d''(\theta_n^*) \xrightarrow{d} d''(\theta_0)$ and $\eta''(\theta_n^*) \xrightarrow{d} \eta''(\theta_0)$ (since d'' and η'' are continuous at θ_0;

see Remark 2.15, p. 53). Furthermore, $\overline{T} \xrightarrow{p} \mathbb{E}T = d'(\theta_0)/\eta'(\theta_0)$ by the law of large numbers and Exercise 23 (p. 53). Finally, by asymptotic normality of the MLE (see Corollary 3.27, p. 83), we know that

$$\sqrt{n}(\widehat{\theta}_n - \theta_0) \xrightarrow{d} N\left(0, \frac{\eta'(\theta_0)}{d''(\theta_0)\eta'(\theta_0) - d'(\theta)\eta''(\theta_0)}\right) = \sqrt{\frac{\eta'(\theta_0)}{d''(\theta_0)\eta'(\theta_0) - d'(\theta)\eta''(\theta_0)}} Z,$$

for some $Z \sim N(0, 1)$. Combining all of the above with Slutsky's theorem (Theorem 2.26, p. 57) gives

$$[\eta''(\theta_n^*)\overline{T} - d''(\theta_n^*)][\sqrt{n}(\widehat{\theta}_n - \theta_0)]^2 \xrightarrow{d} \frac{d''(\theta_0)\eta'(\theta_0) - \eta''(\theta_0)d'(\theta_0)}{\eta'(\theta_0)} \frac{\eta'(\theta_0)}{d''(\theta_0)\eta'(\theta_0) - d'(\theta)\eta''(\theta_0)} Z^2.$$

In other words, $2(\ell(\widehat{\theta}_n) - \ell(\theta_0)) \xrightarrow{d} \chi_1^2$, since $Z^2 \sim \chi_1^2$, being the square of a standard normal random variable (see Eq. (1.4), in Example 1.29, p. 20). □

Exercise 54 Let $X_1, \ldots, X_n$ be an iid sample from a Poisson distribution with parameter θ. We wish to test $H_0 : \theta = \theta_0$ vs $H_1 : \theta \neq \theta_0$. Find an approximate likelihood ratio test for this pair of hypotheses.

4.3.3.3 Wald Tests
Another idea for building tests for bilateral hypotheses $\{H_0 : \theta = \theta_0, H_1 : \theta \neq \theta_0\}$ is to directly use the technology that we've developed for point estimation in order to construct a test function. Suppose that we have an estimator $\hat{\theta}$ of θ. Then, we could compare the null value θ_0 to the observed value of the estimator $\hat{\theta}(X_1, \ldots, X_n)$. If these are separated by a "significant" distance, then it is clear that we should reject $H_0 : \theta = \theta_0$ in favour of $H_1 : \theta \neq \theta_0$. Clearly this distance cannot be expressed in absolute terms, as it needs to take into account the variability of $\hat{\theta}$; so one idea is to express this distance in terms of the variance of $\hat{\theta}$. This leads to a test statistic of the form:

$$T = \frac{(\hat{\theta} - \theta_0)^2}{Var(\hat{\theta})}.$$

and then the test function will be $\delta(X_1, \ldots, X_n) = \mathbf{1}\{T > Q\}$. The critical value Q will of course be chosen in order to ensure that the level of the test is α, in other words we ask that $\mathbb{P}_{\theta_0}[T > Q] = \alpha$. The problem is that $Var(\hat{\theta})$ is typically unknown, and so an estimator $\widehat{Var}(\hat{\theta})$ must be used instead. Using such an estimator, we obtain what is called a *Wald test*.

Definition 4.25 (Wald Test)

Let $X_1, \ldots, X_n \overset{iid}{\sim} f(\cdot; \theta)$ and $\hat{\theta}$ be an estimator of θ based on the sample $X_1, \ldots, X_n$. A Wald test for the bilateral hypothesis pair $\{H_0 : \theta = \theta_0, H_1 :$

$\theta \neq \theta_0\}$ at level α is a test with test function

$$\delta(X_1, \ldots, X_n) = \mathbf{1}\left\{ \frac{(\hat{\theta} - \theta_0)^2}{\widehat{Var}(\hat{\theta})} > Q \right\}$$

where $\mathbb{P}_{\theta_0}\left[\frac{(\hat{\theta}-\theta_0)^2}{\widehat{Var}(\hat{\theta})} > Q \right] = \alpha$, provided such a Q exists.

If $\hat{\theta}$ is taken to be a maximum likelihood estimator of θ, then we have seen (Remark 3.29, p. 85), Exercise 36 (p. 85) that the asymptotic variance equals

$$\frac{1}{n} \frac{[\eta'(\theta_0)]}{d''(\theta_0)\eta'(\theta_0) - d'(\theta_0)\eta''(\theta_0)} = \frac{1}{nI(\theta)} = \frac{1}{nJ(\theta)}.$$

Therefore, we could use

$$\widehat{J}_n = nJ(\hat{\theta}_n) = n\frac{d''(\hat{\theta}_n)\eta'(\hat{\theta}_n) - d'(\hat{\theta}_n)\eta''(\hat{\theta}_n)}{[\eta'(\hat{\theta}_n)]}$$

instead of $\widehat{Var}^{-1}(\hat{\theta})$. When we use $\hat{\theta}$ as the estimator and $\widehat{J}_n$ instead of $\widehat{Var}^{-1}(\hat{\theta})$ in a test of this type, then we get the so-called *likelihood-based Wald test*.

4.3.3.4 Approximate Critical Values for Likelihood-Based Wald Tests

As was the case with likelihood ratio tests, we will rarely be able to find the critical value Q exactly. Instead, we will need an asymptotic approximation with respect to n. For a Wald test based on the likelihood estimator, this approximation can be easily obtained by using our results on the asymptotic distribution of the maximum likelihood estimator. We will consider, as usual, the case of a one-parameter exponential family. The assumptions that we will make are the same as those made when considering approximate critical values for likelihood ratio tests.

Theorem 4.26 (Approximate Critical Values for Wald Tests) *Let $X_1, \ldots, X_n$ be an iid sample from a distribution with density (or mass function) $f(x; \theta)$ which belongs to a non-degenerate one-parameter exponential family,*

$$f(x; \theta) = \exp\{\eta(\theta)T(x) - d(\theta) + S(x)\}, \qquad x \in \mathcal{X}, \theta \in \Theta$$

Assume that:
1. *The parameter space $\Theta \subset \mathbb{R}$ is an open set.*
2. *The function $\eta(\cdot)$ is a twice continuously differentiable bijection between Θ and $\Phi = \eta(\Theta)$ with non-vanishing derivative.*

Let $\hat{\theta}_n$ be the maximum likelihood estimator of θ, and $\widehat{J}_n = nJ(\hat{\theta}_n) = n\frac{d''(\hat{\theta}_n)\eta'(\hat{\theta}_n)-d'(\hat{\theta}_n)\eta''(\hat{\theta}_n)}{[\eta'(\hat{\theta}_n)]}$. Take $\theta_0 \in \Theta$ to be some fixed element of the parameter space. Then,

$$\widehat{J}_n(\hat{\theta}_n - \theta_0)^2 \xrightarrow{d} \chi_1^2,$$

whenever $\{H_0 : \theta = \theta_0\}$ is true.

▶ **Remark 4.27 (Approximate Critical Values for Wald Tests)** The result can now be used in order to determine the right critical value for a Wald test at level α. The Wald test function at level α, say $\mathbf{1}\{\widehat{J}_n(\hat{\theta}_n - \theta_0)^2 > Q\}$, will be approximately (for n being large) equivalent to the test function

$$\mathbf{1}\left\{\widehat{J}_n(\hat{\theta}_n - \theta_0)^2 > \chi_{1,1-\alpha}^2\right\},$$

where $\chi_{1,1-\alpha}^2$ denotes the $(1-\alpha)$-quantile of the χ_1^2 distribution. In other words, for large n, the approximate critical value should be $Q \approx \chi_{1,1-\alpha}^2$.

Proof of Theorem 4.26 Under the conditions of the theorem, and when $\{H_0 : \theta = \theta_0\}$ is true, we may invoke Corollary 3.27 (p. 83) to obtain

$$\sqrt{n}(\hat{\theta}_n - \theta_0) \xrightarrow{d} N\left(0, \frac{[\eta'(\theta_0)]}{d''(\theta_0)\eta'(\theta_0) - d'(\theta_0)\eta''(\theta_0)}\right). \qquad (4.3)$$

Now we may calculate that

$$\frac{1}{n}\widehat{J}_n = \frac{d''(\hat{\theta}_n)\eta'(\hat{\theta}_n) - d'(\hat{\theta}_n)\eta''(\hat{\theta}_n)}{[\eta'(\hat{\theta}_n)]}.$$

By our smoothness assumptions on η (and their ramifications on the smoothness of d, see Remark 2.15, p. 53), the right-hand side above is a continuous function of $\hat{\theta}_n$. Since $\hat{\theta}_n$ is consistent, we may apply the continuous mapping theorem Theorem 2.25 (p. 57) to conclude that

$$\frac{1}{n}\widehat{J}_n \xrightarrow{p} \frac{d''(\theta_0)\eta'(\theta_0) - d'(\theta_0)\eta''(\theta_0)}{[\eta'(\theta_0)]}. \qquad (4.4)$$

Combining (4.3) with (4.4), and using Slutsky's theorem (Theorem 2.26, p. 57) we conclude that

$$\sqrt{\widehat{J}_n}(\hat{\theta}_n - \theta_0) \xrightarrow{d} N(0, 1).$$

Now we may take the square of the left-hand side, and use the continuous mapping theorem (Theorem 2.25, p. 57) to conclude that

$$\left[\sqrt{\widehat{J}_n}(\hat{\theta}_n - \theta_0)\right]^2 = \widehat{J}_n(\hat{\theta}_n - \theta_0)^2 \xrightarrow{d} \chi_1^2$$

because we have seen that the square of a standard normal random variable has the χ_1^2 distribution (see Eq. (1.4), in Example 1.29, p. 20).   □

Exercise 55 Let $X_1, \ldots, X_n$ be an iid $N(0, \sigma^2)$ sample, where the variance σ^2 is unknown. Construct an approximate Wald test (at level α) for the hypothesis pair $H_0 : \sigma^2 = \sigma_0^2$ vs $H_1 : \sigma^2 \neq \sigma_0^2$, for $\sigma_0^2 > 0$ fixed. Compare this test with the corresponding likelihood ratio test.

Exercise 56 Let $X_1, \ldots, X_n$ be iid Bernoulli random variables with unknown parameter p. Construct an approximate Wald test (at level α) for the hypothesis pair $H_0 : p = p_0$ vs $H_1 : p \neq p_0$ for $p_0 \in (0, 1)$ fixed. Compare this test with the corresponding likelihood ratio test.

4.4   The p-Value

We saw that, in the Neyman–Pearson framework, we first need to select a significance level α, and then construct our testing procedure in a way that maximises power, while preserving the level α. This yields a reasonable mathematical theory that can be considered to adequately address the hypothesis testing problem.

There are, nevertheless, two non-negligible weak points when it comes to practical problems. They can be loosely stated as follows:

1. It is not always clear a priori what the "right" significance level is. Should we take $\alpha = 0.05$, or should we take $\alpha = 0.04$? It is the scientist who should suggest what the "right" significance level is, and then the mathematician gives the test function. But what if the scientist does not really know what the precise level should be, or if two different scientists suggest two different levels? This can be an issue because it might be that, for the same data, picking $\alpha = 0.05$ could result in H_0 being rejected, while picking $\alpha = 0.04$ could result in H_0 not being rejected.

2. Suppose we are somehow able to pick a precise level α, so that we have bypassed the problem stated above. Once the level is set, we use the optimal test (if available), and then for our given data set we make a decision. Suppose we reject H_0 at the level α. The problem now is that we have no clear indication of how comfortable or how marginal our decision was. For instance, would our decision have been different, had we selected a slightly smaller α?

Fisher popularised an approach that can be thought of as the dual of the Neyman–Pearson approach, and that provides a means to tackle these two issues. The idea is that, instead of making an binary statement (i.e. $\delta = 0$ or $\delta = 1$), we define a continuous measure of how strong the evidence in the data is against the null hypothesis. This measure is called the *p*-value.

Definition 4.28 (*p*-Value)

Let $X_1, \ldots, X_n \overset{iid}{\sim} f(\cdot; \theta)$ and $H_0 : \theta \in \Theta_0$ be a null hypothesis that is of one of the three following forms:

$$\{H_0 : \theta = \theta_0\} \qquad \text{or} \qquad \{H_0 : \theta \le \theta_0\} \qquad \text{or} \qquad \{H_0 : \theta \ge \theta_0\}.$$

Let δ_α be a test function for H_0, of one of the two following forms:

$$\delta_\alpha(X_1, \ldots, X_n) := \mathbf{1}\{T(X_1, \ldots, X_n) > q_{1-\alpha}\} \quad \text{or} \quad \delta_\alpha(X_1, \ldots, X_n) := \mathbf{1}\{T(X_1, \ldots, X_n) \le q_\alpha\},$$

where T is some test statistic, and q_z is the z-quantile of the distribution $G_0(t) = \mathbb{P}_{\theta_0}[T(X_1, \ldots, X_n) \le t]$. Then, we define

$$p(X_1, \ldots, X_n) := \inf\{\alpha \in (0, 1) : \delta_\alpha(X_1, \ldots, X_n) = 1\}.$$

to be the *p*-value.

▶ **Remark 4.29** Notice that, in all the tests that we have seen, the test function always reduces to one of the two forms mentioned in the definition above, though sometimes perhaps approximately as $n \to \infty$.

In other words, the *p*-value is a random variable that tells us which is the smallest significance level α for which our testing method would reject the null hypothesis H_0 on the basis of the sample $X_1, \ldots, X_n$. Why does this quantity have any relevance? Because it gives us a measure of how stable our decision is under perturbations of a given level α: if the *p*-value is very small, then this means that we reject H_0 even if we are very strict and impose a rather small α (i.e. very small probability of type I error). If the *p*-value is relatively large, this means that we would only have rejected H_0 if we were willing to tolerate a high probability of type I error. How small should the *p*-value be in order to decide that we have rejected ? The answer is left up to the scientist, who can decide depending on his/her deeper knowledge of the experiment at hand. Notice that this approach gives a solution to the problems (1) and (2) outlined above.

The definition of the *p*-value seems a bit complicated, and it is natural to wonder whether it is possible to actually calculate it in concrete examples. This is indeed the case, when the null hypothesis is of one of the forms we have considered thus far; and, in fact, the calculation is quite easy:

Lemma 4.30 (Calculation of p-Values) *In the setup given in Definition 4.28, we have:*

1. If δ_α is of the form $\delta_\alpha(X_1, \ldots, X_n) := \mathbf{1}\{T(X_1, \ldots, X_n) > q_{1-\alpha}\}$, then

$$p(X_1, \ldots, X_n) = 1 - G_0(T(X_1, \ldots, X_n))$$

2. If δ_α is of the form $\delta_\alpha(X_1, \ldots, X_n) := \mathbf{1}\{T(X_1, \ldots, X_n) \leq q_\alpha\}$, then

$$p(X_1, \ldots, X_n) = G_0(T(X_1, \ldots, X_n))$$

▶ **Remark 4.31 (Interpreting p-Values)** The Lemma gives us a further way of understanding p-values. Let's concentrate on case (1), where we reject for large values of T. Notice that $1 - G_0(T(X_1, \ldots, X_n))$ equals the probability of observing something as large, or even larger than what we observed, when H_0 is true. Therefore, when the p-value is small, we have in fact observed something that would be very improbable/unusual if H_0 were indeed true. So we expect that H_0 is false. A common mistake is to interpret the p-value as the *probability that H_0 is true*. This is wrong, and in fact does not even make sense, because the parameter θ is not a random variable.

Proof of Lemma 4.30 It suffices to prove (1), as (2) is proven directly analogously. In the setting (1), we can use the fact that G_0 is non-decreasing to write:

$$\delta_\alpha(X_1, \ldots, X_n) = 1 \implies T(X_1, \ldots, X_n) > q_{1-\alpha} \implies G_0(T(X_1, \ldots, X_n)) \geq G_0(q_{1-\alpha})$$
$$\implies G_0(T(X_1, \ldots, X_n)) \geq 1 - \alpha \implies \alpha \geq 1 - G_0(T(X_1, \ldots, X_n)).$$

It follows that $\inf\{\alpha \in (0, 1) : \delta_\alpha(X_1, \ldots, X_n) = 1\} = 1 - G_0(T(X_1, \ldots, X_n))$, and the proof is complete. $\qquad\qquad\qquad\qquad\qquad\qquad\qquad\qquad\qquad\qquad\qquad\qquad\qquad\quad$ $\square$

Example 4.32

Let $X_1, \ldots, X_n \overset{iid}{\sim} N(\mu, 1)$ and consider the hypothesis pair:

$$H_0 : \mu = 0 \qquad \text{vs} \qquad H_1 : \mu \neq 0$$

We recall (see Example 4.21, p. 115) that the likelihood ratio test for this pair is given by:

$$\delta(X_1, \ldots, X_n) = \mathbf{1}\left\{ \left(\frac{\bar{X}}{1/\sqrt{n}} \right)^2 > \chi^2_{1,1-\alpha} \right\},$$

where $\chi^2_{1,1-\alpha}$ is the $1 - \alpha$ quantile of the χ^2_1 distribution. Notice, therefore, that this test statistic conforms to the setup given in Definition 4.28. We may thus define the corresponding p-value as

$$p(X_1, \ldots, X_n) = 1 - G_{\chi^2_1}\left(n\bar{X}^2 \right),$$

where $G_{\chi_1^2}$ denotes the CDF of the χ_1^2 distribution. Observe that when $\bar{X}$ is at a large distance from 0, then the *p*-value will be small. In fact, the *p*-value is monotonically decreasing in $\bar{X}$ (note that $G_{\chi_1^2}$ is a monotonically increasing function from $(0, \infty)$ to $(0, 1)$ because the density of a χ_1^2 is strictly positive over the entire interval $(0, \infty)$—see Definition 1.16, p. 13). $\qquad\square$

One might finally ask: is there any link between Fisher's and Neyman & Pearson's approach to hypothesis tests? In the case where $G_0(t)$ is strictly monotonic,[2] there is a particularly simple and elegant connection:

> **Corollary 4.33** *In the setup given in Definition 4.28, let $\alpha_0 \in (0, 1)$ and assume that G_0 is continuous and strictly increasing. If we define a test function*
>
> $$\psi(X_1, \ldots, X_n) := \mathbf{1}\{p(X_1, \ldots, X_n) \le \alpha_0\},$$
>
> *then $\psi(X_1, \ldots, X_n) = \delta_{\alpha_0}(X_1, \ldots, X_n)$. In other words, if we reject the null whenever the p-value is smaller than α_0, then our test reduces to δ_{α_0}.*

Proof Without loss of generality, we assume that we are in the setup where the *p*-value corresponds to a statistic of the form $\delta_\alpha(X_1, \ldots, X_n) := \mathbf{1}\{T(X_1, \ldots, X_n) > q_{1-\alpha}\}$. Now, observe that, using Lemma 4.30, and we have:

$$p(X_1, \ldots, X_n) < \alpha_0 \iff 1 - G_0(T(X_1, \ldots, X_n)) < \alpha_0 \iff G_0(T(X_1, \ldots, X_n)) > 1 - \alpha_0.$$

Under our assumptions, G_0^{-1} exists and is strictly increasing. Applying it to both sides of the last inequality yields:

$$p(X_1, \ldots, X_n) < \alpha_0 \iff T(X_1, \ldots, X_n) > \underbrace{G_0^{-1}(1 - \alpha_0)}_{=q_{1-\alpha_0}} \iff \delta(X_1, \ldots, X_n) = 1.$$

$\qquad\square$

It follows that the *p*-value is a versatile tool: reporting a p-value solves some of the problems that we mentioned earlier in this paragraph. Still, even when we report a *p*-value, we can still use it to implement a Neyman–Pearson type test at some level α, simply by rejecting whenever $p < \alpha$.

---

[2]This is not as restrictive as it may sound. A sufficient condition is that the distribution must be of the continuous type, with a probability density function satisfying $g_0(t) > 0$ for all t. This will be true, for example whenever G_0 is a CDF corresponding to a normal, Student, or exponential family distribution. Furthermore, in many examples, we can approximate G_0 for large n by the normal CDF, so the assumption is again approximately satisfied, even if the exact form of G_0 is discrete.

Exercise 57 Let $X_1, \ldots, X_n \overset{iid}{\sim} f(x; \theta)$. Suppose we wish to test $H_0 : \theta = \theta_0$ vs $H_1 : \theta \neq \theta_0$ using the test function δ_α of the form

$$\delta_\alpha(T(X_1, \ldots, X_n)) = \mathbf{1}\{T(X_1, \ldots, X_n) > q_{1-\alpha}\} \text{ or } \delta_\alpha(T(X_1, \ldots, X_n)) = \mathbf{1}\{T(X_1, \ldots, X_n) \leq q_\alpha\},$$

where q_α is the α-quantile of G_0, the CDF of $T(X_1, \ldots, X_n)$ when $\theta = \theta_0$. Assuming that G_0 is continuous, show that, under H_0, the p-value is uniformly distributed on $[0, 1]$.

4.5   On Terminology: Accepting Versus Not Rejecting

From the mathematical perspective, the outcome of a hypothesis test is clear cut: 0 or 1. This means that we decide between the two competing hypotheses, H_0 and H_1. How do we communicate this decision in the context of an application?

In the context of science, competing hypotheses represent competing scientific theories. The null hypothesis represents a scientific assertion. The alternative hypothesis encapsulates how we might expect the assertion to break down.

When the outcome of the test is 0, then the empirical evidence is not sufficient in order to reject the null hypothesis. Does this mean that the evidence actually proves that H_0 is true? No, it merely does not disprove H_0. For this reason, when the outcome is 0, we say that "we do not reject the null hypothesis H_0" instead of saying "we accept the null hypothesis H_0". From a mathematical perspective, we can think of this in the context of necessary and sufficient conditions. If the evidence is such that $\delta = 0$, then a necessary condition for H_0 to hold true (=the data being consistent with H_0) is not violated. This does not *prove validity* of H_0, it merely says that we cannot *disprove validity* of H_0 given the current data set.

On the other hand, when the test results in "1", the interpretation is that the evidence does not support the null hypothesis: the data appear to be incompatible with H_0 (we have something like a counterexample). So, we can say that "we reject the null". But can we actually say that "we accept the alternative"? The alternative was used as a device in order to detect possible departures from the null, by constructing a test function that would be able to detect departures in the direction of the alternative. It was our "best devil's advocate", but it was not necessarily the most viable alternative scientific theory in itself. For this reason, in the context of scientific applications, when $\delta = 1$, we almost always say "we reject the null hypothesis H_0", instead of saying "we accept the alternative hypothesis H_1".

Again, from a mathematical standpoint, things are clear: we decide 0 or 1. But when communicating a mathematical result to scientists, there are pitfalls due to the weaknesses of the verbal presentation of otherwise rigorous mathematical results. The language of mathematics is clear, but the verbal presentation of mathematics will always be less rigorous, and care must be taken.

In summary, the next table presents the recommended way of verbally conveying the result of a hypothesis test:

Mathematical statement	$\delta(X_1, \ldots, X_n) = 1$	$\delta(X_1, \ldots, X_n) = 0$
Verbal statement	We reject the null hypothesis	We do not reject the null hypothesis

Exercise 58 As an example of a situation where one must be careful in phrasing the result of a test, we consider a more complicated scenario. Let (X, Y), be a random vector taking values in $\{1, 2\}^2$. Let $(X_1, Y_1), \ldots, (X_n, Y_n)$ be an iid random sample distributed as (X, Y). We wish to test the hypothesis that X and Y are independent random variables. Let $p_1 = \mathbb{P}(X = 1)$, $p_2 = \mathbb{P}(Y = 1)$ and $p_3 = \mathbb{P}(X = Y = 1)$.

1. Formulate the null and alternative hypotheses in terms of p_1, p_2 and p_3.
2. Find the maximum likelihood estimators of $\widehat{p}_1$, $\widehat{p}_2$ and $\widehat{p}_3$ on the basis of the sample values $(x_1, y_1), \ldots, (x_n, y_n)$ in general, as well as when the null hypothesis is valid.
3. Show that if $p_1 = p_2 = 1/2$ are known, we have a one-parameter exponential family. Test the independence hypothesis in this case, and find the approximate p-value for the following data : $n = 1024$, $n_{11} = 266$, $n_{12} = 231$, $n_{21} = 243$, $n_{22} = 284$ where n_{ij} is the number k such that $X_k = i$ and $Y_k = j$.

Remark : There exists a test for the more general case when p_1, p_2, p_3 are unknown, where the limiting distribution of the test statistic is χ_1^2, but we do not yet have the tools to consider this case rigorously. The test applies also when X takes $k > 1$ different values, and Y takes $l > 1$ different values. The limiting distribution will be $\chi^2_{(k-1)(l-1)}$ in this case.

Confidence Intervals for Model Parameters **5**

Once again, let us zoom out to see the bigger picture: there is a regular parametric family of distributions $\mathcal{F} = \{F_\theta : \theta \in \Theta\}$, where $\Theta \subseteq \mathbb{R}$, which is our model for a certain stochastic phenomenon. We are able to observe n independent and identically distributed outcomes from the phenomenon, say $X_1, \ldots, X_n \overset{iid}{\sim} F_\theta$ generated for a particular choice of $\theta \in \Theta \subseteq \mathbb{R}$; but the precise value $\theta \in \Theta$ that generated them (the *true state of nature*) is unknown to us. With this iid sample at our disposal, we wish to make inferences about θ. So far we have made two kinds of inferences on the true parameter value:

1. *Point Estimation.* Find the exact value of the unknown parameter θ, as accurately as possible.
2. *Hypothesis Testing.* Given two candidate regions Θ_1 and Θ_0 where θ might lie, find optimal ways of deciding in which of the two regions the true θ resides.

In this chapter, we will consider the third important problem of statistical inference, which loosely stated is:

3. *Interval Estimation.* Find an interval of plausible values for θ, in the sense that the interval has a high probability of containing θ.

The essence of the third problem is as follows. We know that an estimator $\hat{\theta}(X_1, \ldots, X_n)$ of θ is a random variable. Therefore, the probability that $\hat{\theta}$ perfectly estimates θ is either low (if $\hat{\theta}$ is a discrete random variable) or even zero (if $\hat{\theta}$ is a continuous random variable). However, if $\hat{\theta}$ is an estimator with a low mean squared error, then we expect that θ cannot be very far from our estimate $\hat{\theta}(X_1, \ldots, X_n)$. Can we use our estimator $\hat{\theta}$ and (approximate) knowledge of its sampling distribution in order to propose an interval that is highly likely to contain the true θ? Such an interval we call a *confidence interval*.

In the next few paragraphs we will define the notion of a confidence interval rigorously, and we will show how we can use our knowledge of point estimation theory in order to construct such intervals. We will then consider the problem of how

© Springer International Publishing Switzerland 2016
V.M. Panaretos, *Statistics for Mathematicians*, Compact Textbooks in Mathematics,
DOI 10.1007/978-3-319-28341-8_5

to define "optimal intervals". To do this, we will use an important duality between interval estimation and hypothesis testing.[1]

5.1   Confidence Intervals and Confidence Levels

Let us begin with the rigorous definition of a confidence interval, and then discuss its elements.

Definition 5.1 (Two-Sided Confidence Interval)

Let $X_1, \ldots, X_n \overset{iid}{\sim} f(x; \theta)$, where $\theta \in \Theta \subseteq \mathbb{R}$, be random sample and $\alpha \in (0, 1)$ be a constant. Let $L(X_1, \ldots, X_n)$ and $U(X_1, \ldots, X_n)$ be two statistics, called the lower limit and upper limit, respectively, such that

$$\inf_{\theta \in \Theta} \mathbb{P}_\theta \Big[L(X_1, \ldots, X_n) \leq \theta \leq U(X_1, \ldots, X_n) \Big] \geq 1 - \alpha.$$

Then, the random interval

$$\Big[L(X_1, \ldots, X_n), \, U(X_1, \ldots, X_n) \Big]$$

is called a two-sided confidence interval for θ with confidence level $(1 - \alpha)$.

Since anything we do will depend on our sample $X_1, \ldots, X_n$, any candidate interval we propose will in fact be a random interval that will take different values for different realisations of our sample. In order to be able to construct this random interval, its endpoints L and U will be statistics constructed from our sample.

For the interval to truly be a likely region for the true parameter θ, we ask that the probability of the event $\{L \leq \theta \leq U\}$ be at least as large as $1 - \alpha$, whatever the true value of θ may be[2] for some small probability α. There are situations where we are more interested in giving a lower or upper confidence bound on the true value of a parameter θ. In these cases, instead of using a two-sided confidence interval as defined in Definition 5.1, we use the notion of a one-sided interval.

Definition 5.2 (One-Sided Confidence Interval)

Let $X_1, \ldots, X_n \overset{iid}{\sim} f(x; \theta)$, where $\theta \in \Theta \subseteq \mathbb{R}$, be random sample and $\alpha \in (0, 1)$ be a constant. Let $L(X_1, \ldots, X_n)$ be a statistic such that

$$\inf_{\theta \in \Theta} \mathbb{P}_\theta \Big[L(X_1, \ldots, X_n) \leq \theta \Big] \geq 1 - \alpha.$$

---

[1] Note that the problem "use the data to decide if the region Θ_0 contains θ" is in some sense dual to the question "use the data to find a region that is highly likely to contain θ".

[2] Since this probability obviously depends on the true value of θ!

Then, the random interval

$$\left[L(X_1, \ldots, X_n) , +\infty \right)$$

is called a left-sided confidence interval for θ with confidence level $(1 - \alpha)$. Analogously, if $U(X_1, \ldots, X_n)$ is a statistic such that

$$\inf_{\theta \in \Theta} \mathbb{P}_\theta \left[U(X_1, \ldots, X_n) \geq \theta \right] \geq 1 - \alpha,$$

then the random interval

$$\left(-\infty , U(X_1, \ldots, X_n) \right]$$

is called a right-sided confidence interval for θ with confidence level $(1 - \alpha)$.

We now illustrate many essential features of confidence intervals within the framework of the following prototypical example.

Example 5.3 (Confidence Interval for the Mean of a Normal Distribution)

Let $X_1, \ldots, X_n \overset{iid}{\sim} N(\mu, \sigma^2)$, where μ is unknown and σ^2 is known. We wish to construct a two-sided interval for μ. We begin by observing that by Lemma (1.32, p. 22) we have:

$$\frac{\bar{X} - \mu}{\sigma / \sqrt{n}} \sim N(0, 1).$$

Therefore, if $z_{\frac{\alpha}{2}}$ and $z_{1-\frac{\alpha}{2}}$ are the $\alpha/2$ and $1 - \alpha/2$ quantiles (respectively) of the $N(0, 1)$ distribution, we must have:

$$\mathbb{P}\left[z_{\frac{\alpha}{2}} \leq \frac{\bar{X} - \mu}{\sigma / \sqrt{n}} \leq z_{1-\frac{\alpha}{2}} \right] = 1 - \alpha.$$

Now, let us manipulate the expression inside the probability:

$$\mathbb{P}\left[z_{\frac{\alpha}{2}} \leq \frac{\bar{X} - \mu}{\sigma / \sqrt{n}} \leq z_{1-\frac{\alpha}{2}} \right] = 1 - \alpha$$

$$\Longleftrightarrow \ \mathbb{P}\left[z_{\frac{\alpha}{2}} \frac{\sigma}{\sqrt{n}} \leq \bar{X} - \mu \leq z_{1-\frac{\alpha}{2}} \frac{\sigma}{\sqrt{n}} \right] = 1 - \alpha$$

$$\Longleftrightarrow \ \mathbb{P}\left[-\bar{X} + z_{\frac{\alpha}{2}} \frac{\sigma}{\sqrt{n}} \leq -\mu \leq -\bar{X} + z_{1-\frac{\alpha}{2}} \frac{\sigma}{\sqrt{n}} \right] = 1 - \alpha$$

$$\Longleftrightarrow \; \mathbb{P}\left[\bar{X} - z_{\frac{\alpha}{2}} \frac{\sigma}{\sqrt{n}} \geq \mu \geq \bar{X} - z_{1-\frac{\alpha}{2}} \frac{\sigma}{\sqrt{n}}\right] = 1 - \alpha$$

$$\Longleftrightarrow \; \mathbb{P}\left[\bar{X} - z_{1-\frac{\alpha}{2}} \frac{\sigma}{\sqrt{n}} \leq \mu \leq \bar{X} - z_{\frac{\alpha}{2}} \frac{\sigma}{\sqrt{n}}\right] = 1 - \alpha.$$

The above equality is true whatever the true value of $\mu \in \mathbb{R}$ may be. It follows that if we set

$$L(X_1, \ldots, X_n) = \bar{X} - z_{1-\frac{\alpha}{2}} \frac{\sigma}{\sqrt{n}} \qquad \& \qquad U(X_1, \ldots, X_n) = \bar{X} - z_{\frac{\alpha}{2}} \frac{\sigma}{\sqrt{n}}$$

then the interval $[L, U]$ is a confidence interval with confidence level $1 - \alpha$. Because the density of an $N(0, 1)$ distribution is symmetric, we have that $z_{\frac{\alpha}{2}} = -z_{1-\frac{\alpha}{2}}$. So our $(1 - \alpha)$-confidence interval may be written as

$$\left[\underbrace{\bar{X} - z_{1-\frac{\alpha}{2}} \frac{\sigma}{\sqrt{n}}}_{L(X_1,\ldots,X_n)} \quad , \quad \underbrace{\bar{X} + z_{1-\frac{\alpha}{2}} \frac{\sigma}{\sqrt{n}}}_{U(X_1,\ldots,X_n)}\right] \tag{5.1}$$

For brevity, we sometimes represent the endpoints of the interval as $\bar{X} \pm z_{1-\frac{\alpha}{2}} \frac{\sigma}{\sqrt{n}}$. Notice that the confidence interval is centred around the maximum likelihood estimator of μ. It says that our plausible region is the MLE, plus or minus a constant times the standard deviation of the MLE (since σ^2/n is the variance of the MLE $\bar{X}$). The constant is chosen in order to have confidence level $1 - \alpha$.

We can also make some more observations. The length of the confidence interval (equal to $2z_{1-\alpha/2}\sigma/\sqrt{n}$) depends on σ^2, n and α. The parameter σ^2 is beyond our control, since it is the variance of the underlying $N(\mu, \sigma^2)$ distribution. The two parameters that we are able to control are the sample size n and the confidence level $1 - \alpha$. Increasing n re-scales the length by $1/\sqrt{n}$. So, for example, if we want to make the interval ten times shorter, we need to take a sample size that is 100 times larger. On the other hand, decreasing α (increasing the confidence $1 - \alpha$) increases the length of the interval: the more confident we want to be in our interval, the longer the interval will be (notice that the length of the interval tends to ∞ as $\alpha \to 0$).

We may also ask how to construct one-sided confidence intervals, in case we are interested in lower or upper bounds for the parameter μ. Let us consider the problem of finding a right-sided confidence interval. Using the fact that $\frac{\bar{X}-\mu}{\sigma/\sqrt{n}} \sim N(0, 1)$, we may write

$$\mathbb{P}\left[\frac{\bar{X} - \mu}{\sigma/\sqrt{n}} \geq z_\alpha\right] = 1 - \alpha.$$

This can be manipulated to yield

$$\mathbb{P}\left[\bar{X} + z_{1-\alpha} \frac{\sigma}{\sqrt{n}} \geq \mu\right] = 1 - \alpha,$$

and so the interval

$$\left(-\infty \quad , \quad \bar{X} + z_{1-\alpha} \frac{\sigma}{\sqrt{n}}\right]$$

is a right-sided confidence interval with confidence level $1 - \alpha$. Similarly, we can show that a left-sided $(1 - \alpha)$-confidence interval is given by

$$\left[\; \bar{X} - z_{1-\alpha} \frac{\sigma}{\sqrt{n}} \; , \; +\infty \; \right).$$

In summary:

Confidence $1 - \alpha$	$L(X_1, \ldots, X_n)$	$U(X_1, \ldots, X_n)$
Two-sided	$\bar{X} - z_{1-\alpha/2} \dfrac{\sigma}{\sqrt{n}}$	$\bar{X} + z_{1-\alpha/2} \dfrac{\sigma}{\sqrt{n}}$
Left-sided	$\bar{X} - z_{1-\alpha} \dfrac{\sigma}{\sqrt{n}}$	$+\infty$
Right-sided	$-\infty$	$\bar{X} + z_{1-\alpha} \dfrac{\sigma}{\sqrt{n}}$

$\square$

Exercise 59 (Normal Case, Unknown Variance) Let $X_1, \ldots, X_n \overset{iid}{\sim} N(\mu, \sigma^2)$, where both μ and σ^2 are unknown. Let $S^2 = \sum_{i=1}^{n}(X_i - \bar{X})^2/(n-1)$, and $t_{\{k,\alpha\}}$ be the α-quantile of Student's t_k distribution (with k degrees of freedom). Prove that the confidence intervals given by the following table are $(1-\alpha)$-confidence intervals for the mean μ.

Confidence $1 - \alpha$	$L(X_1, \ldots, X_n)$	$U(X_1, \ldots, X_n)$
Two-sided	$\bar{X} - t_{\{n-1,1-\alpha/2\}} \dfrac{S}{\sqrt{n}}$	$\bar{X} + t_{\{n-1,1-\alpha/2\}} \dfrac{S}{\sqrt{n}}$
Left-sided	$\bar{X} - t_{\{n-1,1-\alpha\}} \dfrac{S}{\sqrt{n}}$	$+\infty$
Right-sided	$-\infty$	$\bar{X} + t_{\{n-1,1-\alpha\}} \dfrac{S}{\sqrt{n}}$

Exercise 60 (Optimal Choice of Quantiles) In order to construct the two-sided confidence interval for the mean of a normal distribution (known variance) in Example 5.3, we chose $z_{\alpha/2}$ and $z_{1-\alpha/2}$ as the quantiles to base the interval on. One can wonder why not choose $z_{\alpha/3}$ and $z_{1-2\alpha/3}$, for example. It's true that a more natural choice of interval is a symmetric interval, but here is a further reason why :

1. Let $Z \sim N(0, 1)$ and $\alpha \in (0, 1)$. Show that the interval $I = [L, U]$ of minimal length such that $\mathbb{P}(I \ni Z) \geq 1 - \alpha$ is given by the choice $L = z_{\alpha/2}$ and $U = z_{1-\alpha/2}$.
2. Let $X_1, \ldots, X_n \overset{iid}{\sim} N(\mu, \sigma^2)$ where σ^2 is known. Find the interval $I_n = [A_n, B_n]$ of smallest length such that $\mathbb{P}(I_n \ni \mu) \geq 1 - \alpha$.
3. Can we generalise this result to the case of unknown variance? Or even to distributions other than the normal distribution?

Exercise 61 (Difference Between Means) Let $X_1, \ldots, X_n \overset{\text{iid}}{\sim} N(\mu_X, \sigma^2)$ and $Y_1, \ldots, Y_n \overset{\text{iid}}{\sim} N(\mu_Y, \sigma^2)$ be two independent samples, where μ_X, μ_Y and σ^2 are all unknown. Construct a two-sided confidence interval for the parameter $\theta = \mu_X - \mu_Y$ with confidence level $1 - \alpha$.

5.2   Pivots and Approximate Pivots

It seems that the construction of confidence intervals is quite straightforward and indeed transparent in the case of the mean parameter of a normal distribution. However, it also seems that the way we proceeded in our construction was rather ad-hoc, and indeed specific to that particular case. How does this example make us any wiser in terms of constructing confidence intervals in more general situations? We need to find general methods of constructing such intervals. The crucial step in Example 5.3 was exploiting the fact that

$$\frac{\bar{X} - \mu}{\sigma/\sqrt{n}} \sim N(0, 1).$$

This allowed us to write the probability statement

$$\mathbb{P}\left[z_{\alpha/2} \leq \frac{\bar{X} - \mu}{\sigma/\sqrt{n}} \leq z_{1-\alpha/2} \right] = 1 - \alpha$$

which was valid for any value of μ. We were then able to manipulate the argument of the probability to get our interval. The reason this worked was that $\frac{\bar{X}-\mu}{\sigma/\sqrt{n}}$ constitutes what we call a *pivot*.

Definition 5.4 (Pivot)

Let $X_1, \ldots, X_n \overset{iid}{\sim} f(x; \theta)$. A function

$$g : \mathcal{X}^n \times \Theta \to \mathbb{R}$$

is called a pivot if:
1. $\theta \mapsto g(x_1, \ldots, x_n, \theta)$ is continuous for all $(x_1, \ldots, x_n) \in \mathcal{X}^n$.
2. $\mathbb{P}[g(X_1, \ldots, X_n, \theta) \leq x]$ does not depend on θ.

▶ **Remark 5.5** In other words, a pivot $g(X_1, \ldots, X_n, \theta)$ is a function of the sample and the parameter, but its distribution is not a function of the parameter. Notice that, by its very definition, a pivot is <u>not</u> a statistic: it depends on the unknown parameter! The continuity requirement will become clear soon.

If we are able to find a pivot for θ, whose distribution is known, then we are able to find quantiles q_1 and q_2 such that

$$\mathbb{P}[q_1 \leq g(X_1, \ldots, X_n, \theta) \leq q_2] = 1 - \alpha.$$

If g is of a form that allows us to manipulate the inequality inside the probability (similarly to Example 5.3), then we are able to obtain an explicit confidence interval. Still, though, even if we cannot manipulate the expression, we can numerically try to determine the set

$$\{\theta \in \Theta : q_1 \leq g(X_1, \ldots, X_n, \theta) \leq q_2\}$$

and retain this set as our confidence interval. Notice that under our continuity assumption (2) on g, this set may be an interval or a union of intervals depending on the behaviour of g. A sufficient condition to obtain a single interval is to ask that g be monotone in θ. But this is not a necessary condition, of course. In practice, the pivots that we will encounter will typically give us intervals rather than unions of intervals.

Once we have a pivot whose distribution is known, then we are able to construct confidence intervals. However, there are two challenges that we now face:
1. How can we find pivots in general?
2. How can we determine the distribution of a pivot?

The determination of a pivot (and its distribution) depends upon the particular probability distribution, and also on which parameter of the distribution we wish to construct a confidence interval for. Thus, there is no single "explicit formula", and pivots are constructed on a case-by-case basis. Nevertheless, it turns out that, often we can answer both questions (1) and (2) with a general "explicit formula" by settling for what is called an *approximate pivot*. This means that it may not be a pivot for a finite n, but gradually satisfy the assumptions of a pivot as $n \to \infty$.

Definition 5.6 (Approximate Pivot)

Let $X_1, \ldots, X_n \overset{iid}{\sim} f(x; \theta)$. A function

$$g : \mathcal{X}^n \times \Theta \to \mathbb{R}$$

is called an approximate pivot if:
1. For all $n \in \mathbb{N}$, $\theta \mapsto g(x_1, \ldots, x_n, \theta)$ is continuous for all $(x_1, \ldots, x_n) \in \mathcal{X}^n$.
2. We have

$$g(X_1, \ldots, X_n, \theta) \overset{d}{\longrightarrow} Y$$

where Y is a random variable whose distribution does not depend on θ.

If we know the asymptotic distribution of an approximate pivot, we may construct an approximate confidence interval. How? Assume that Y is a continuous random variable. If we take q_1 and q_2 to be quantiles of F_Y such that

$$\mathbb{P}[q_1 \leq Y \leq q_2] = 1 - \alpha,$$

then we have

$$g(X_1, \ldots, X_n, \theta) \xrightarrow{d} Y \implies \mathbb{P}[q_1 \leq g(X_1, \ldots, X_n, \theta) \leq q_2] \xrightarrow{n \to \infty} 1 - \alpha.$$

We can therefore use the approximate pivot in order to build an approximate confidence interval.

Example 5.7 (Mean of a General Distribution)

Let $X_1, \ldots, X_n$ be an iid collection of random variables with unknown mean $\mu = \mathbb{E}[X]$ and unknown variance $\mathbb{E}[(X_1 - \mu)^2] = \sigma^2 < \infty$. Suppose we wish to find an approximate pivot in order to construct a $(1 - \alpha)$-confidence interval for μ. We remark that:

- By the central limit theorem (Theorem 2.23, p. 56), we have $\sqrt{n}(\bar{X} - \mu) \xrightarrow{d} N(0, \sigma^2)$.
- By the strong law of large numbers (see Remark 2.22, p. 56), $S_n^2 = \sum_{i=1}^{n}(X_i - \mu)^2/(n-1) \xrightarrow{p} \sigma^2$. Indeed, $U_n^2 = \sum_{i=1}^{n}(X_i - \mu)^2/(n-1) \xrightarrow{p} \sigma^2$ and $U_n^2 - S_n^2 = n(n-1)^{-1}(\bar{X} - \mu)^2 \xrightarrow{p} 0$. Combining the two facts provided above, we may use Slutsky's theorem (Theorem 2.26, p. 57) to conclude that

$$g(X_1, \ldots, X_n, \mu) = \frac{\bar{X} - \mu}{S/\sqrt{n}} \xrightarrow{d} Y \sim N(0, 1).$$

so that we have found an approximate pivot. Mimicking the manipulations carried out in Exercise (5.3, p. 133), we have that:

$$
\begin{aligned}
\mathbb{P}\left[\bar{X} - z_{1-\frac{\alpha}{2}} \frac{S}{\sqrt{n}} \leq \mu \leq \bar{X} - z_{\frac{\alpha}{2}} \frac{S}{\sqrt{n}}\right] &= \mathbb{P}[z_{\alpha/2} \leq \frac{\bar{X} - \mu}{S/\sqrt{n}} \leq z_{1-\alpha/2}] \\
&= \mathbb{P}[z_{\alpha/2} \leq g(X_1, \ldots, X_n, \mu) \leq z_{1-\alpha/2}] \\
&\xrightarrow{n \to \infty} \mathbb{P}[z_{\alpha/2} \leq Y \leq z_{1-\alpha/2}] = 1 - \alpha.
\end{aligned}
$$

It follows that the interval $\bar{X} \pm z_{1-\frac{\alpha}{2}} \frac{S}{\sqrt{n}}$ is approximately, for large n, a two-sided $(1 - \alpha)$-confidence interval for μ. By similar arguments, we may construct one-sided confidence intervals. The results can be summarised in the following table:

Approximate Confidence $1 - \alpha$	$L(X_1, \ldots, X_n)$	$U(X_1, \ldots, X_n)$
Two-sided	$\bar{X} - z_{1-\alpha/2}\dfrac{S}{\sqrt{n}}$	$\bar{X} + z_{1-\alpha/2}\dfrac{S}{\sqrt{n}}$
Left-sided	$\bar{X} - z_{1-\alpha}\dfrac{S}{\sqrt{n}}$	$+\infty$
Right-sided	$-\infty$	$\bar{X} + z_{1-\alpha}\dfrac{S}{\sqrt{n}}$

□

Of course, in general we will be interested in parameters other than just the mean, so this example is rather special. In the next section we shall consider two ways of constructing approximate pivots in one-parameter exponential families.

Exercise 62 Combining the reasoning in Example 5.7 and Example 5.3 (p. 133), to show that if $T_k \sim \mathbf{t}_k$, then $T_k \xrightarrow{d} Z$ as $k \to \infty$, where $Z \sim N(0, 1)$.

5.2.1 Approximate Pivots in Exponential Families

We have seen thus far that both point estimation and hypothesis testing have some very attractive properties when considering one-parameter exponential families. The problem of interval estimation is no exception. We will see in this paragraph that it is feasible to find approximate pivots for one-parameter exponential families under very mild conditions. We will consider two types of confidence intervals arising from two types of pivots:
1. Wald intervals.
2. Likelihood ratio intervals.

Notice that the names of these two methods highly resemble two methods we saw for constructing hypothesis tests. This is no accident, and we will rigorously investigate this connection in Sect. 5.3 (p. 141). For the moment, we determine the approximate pivots.

5.2.1.1 Wald Pivots

Proposition 5.8 (Wald Approximate Pivots) *Let $X_1, \ldots, X_n$ be an iid sample from a distribution with density (or mass function) $f(x; \theta)$ which belongs to a non-degenerate one-parameter exponential family,*

$$f(x; \theta) = \exp\{\eta(\theta)T(x) - d(\theta) + S(x)\}, \qquad x \in \mathcal{X}, \ \theta \in \Theta$$

Assume that:
1. The parameter space $\Theta \subset \mathbb{R}$ is an open set.

2. *The function $\eta(\cdot)$ is a twice continuously differentiable bijection between Θ and $\Phi = \eta(\Theta)$.*
Let $\hat{\theta}_n$ be the maximum likelihood estimator of θ, and $\widehat{J}_n = nJ(\hat{\theta}_n) = n\frac{d''(\hat{\theta}_n)\eta'(\hat{\theta}_n)-d'(\hat{\theta}_n)\eta''(\hat{\theta}_n)}{\eta'(\hat{\theta}_n)}$. Define

$$g(X_1, \ldots, X_n, \theta) := \widehat{J}_n^{1/2}(\hat{\theta}_n - \theta).$$

Then

$$g(X_1, \ldots, X_n, \theta) \xrightarrow{d} N(0, 1),$$

and so $g(X_1, \ldots, X_n, \theta)$ is an approximate pivot for θ.

Proof The proof is exactly the same as that of Theorem (4.26, p. 122) only this time, instead of θ_0, we write θ. $\square$

Exercise 63 (Wald Approximate Confidence Intervals) Using the same notation as in Proposition 5.8 above, prove that the following table indeed yields approximate $(1 - \alpha)$-confidence intervals for θ:

Approximate Confidence $1 - \alpha$	$L(X_1, \ldots, X_n)$	$U(X_1, \ldots, X_n)$
Two-sided	$\hat{\theta} - z_{1-\alpha/2}\widehat{J}_n^{-1/2}$	$\hat{\theta} + z_{1-\alpha/2}\widehat{J}_n^{-1/2}$
Left-sided	$\hat{\theta} - z_{1-\alpha}\widehat{J}_n^{-1/2}$	$+\infty$
Right-sided	$-\infty$	$\hat{\theta} + z_{1-\alpha}\widehat{J}_n^{-1/2}$

5.2.1.2 Likelihood Ratio Pivots

Proposition 5.9 (LRT Approximate Pivots) *Let $X_1, \ldots, X_n$ be an iid sample from a distribution with density (or mass function) $f(x; \theta)$ which belongs to a non-degenerate one-parameter exponential family,*

$$f(x; \theta) = \exp\{\eta(\theta)T(x) - d(\theta) + S(x)\}, \qquad x \in \mathcal{X}, \theta \in \Theta$$

Assume that:
1. *The parameter space $\Theta \subset \mathbb{R}$ is an open set.*
2. *The function $\eta(\cdot)$ is a twice continuously differentiable bijection between Θ and $\Phi = \eta(\Theta)$.*
Let $\hat{\theta}_n$ be the maximum likelihood estimator of θ, and

$$g(X_1, \ldots, X_n, \theta) = 2(\ell(\hat{\theta}) - \ell(\theta)).$$

Then,

$$g(X_1, \ldots, X_n, \theta) \xrightarrow{d} \chi_1^2,$$

and so $g(X_1, \ldots, X_n, \theta)$ is an approximate pivot for θ.

Proof The proof is exactly the same as that of Theorem (4.23, p. 120) only this time, instead of θ_0, we write θ.   □

Notice that the likelihood ratio approximate pivot $g(X_1, \ldots, X_n, \theta) = 2(\ell(\hat{\theta}) - \ell(\theta))$ is not necessarily of a form that we are able to manipulate in order to get the explicit form of the approximate confidence interval. However, we may numerically find the approximate confidence interval of interest, by determining the set

$$\{\theta \in \Theta : g(X_1, \ldots, X_n, \theta) \leq q_{1-\alpha}(\chi_1^2)\},$$

where $q_{1-\alpha}(\chi_1^2)$ is the $(1 - \alpha)$-quantile of the χ_1^2 distribution.

Exercise 64 (Exact and Approximate Pivots)
1. Let $X_1, \ldots, X_n \overset{iid}{\sim} f(x; \theta)$ and $T_n(X_1, \ldots, X_n)$ be a sufficient statistic that is a continuous random variable. Let $Y_n = F_{T_n}(T_n; \theta)$, where $F_{T_n}(t; \theta) = \mathbb{P}_\theta[T_n \leq t]$ is the sampling distribution function of T_n. Show that $Y_n \sim U(0, 1)$ and thus Y_n is a pivot.
2. How can you use this result to construct a confidence interval for θ, in the case where F_{T_n} is known exactly?
3. Assume that $f(x; \theta) = e^{-(x-\theta)}\mathbf{1}\{x \in [\theta, \infty)\}$ (not an exponential family). Use part (1) and the statistic $T_n = \min\{X_1, \ldots, X_n\}$ to find a confidence interval for θ at confidence level $1 - \alpha$.

5.3   The Duality with Hypothesis Tests

The careful reader may have become suspicious that there is a structural connection lurking between confidence intervals and hypothesis tests, while going through the previous paragraphs. Here are some clues that one might have picked up along the way:
- In interval estimation, we try to find a region that will contain the parameter. In hypothesis testing, we are given a region and asked whether it contains the parameter. It seems like the two problems are dual to each other.
- In hypothesis testing we have the level (the probability of falsely rejecting H_0) which is α. In interval estimation we have the confidence level $1 - \alpha$ (the probability that the interval cover the true parameter). Is there a relationship between the two?

- In hypothesis testing, we constructed likelihood ratio tests and Wald tests for the parameter. In interval estimation, we constructed Wald and likelihood ratio intervals for the parameter.

Could it be that we are looking at the two sides of the same coin? This is indeed the case, and it is now time to make the connection rigorous.

Theorem 5.10 (Duality Theorem) *Let* $X_1, \ldots, X_n \overset{iid}{\sim} f(x; \theta)$ *be a random sample and* $\theta \in \Theta \subseteq \mathbb{R}$.

1. If $[L(X_1, \ldots, X_n), U(X_1, \ldots, X_n)]$ *is a two-sided* $(1 - \alpha)$*-confidence interval for* θ, *then the test with test function*

$$\delta(X_1, \ldots, X_n) = \mathbf{1}\{\theta_0 \notin [L(X_1, \ldots, X_n), U(X_1, \ldots, X_n)]\}$$

is a level α *test of* $\{H_0 : \theta = \theta_0\}$ *against* $\{H_1 : \theta \neq \theta_0\}$.

2. Conversely, suppose that given any $\theta_0 \in \Theta$, $\delta(X_1, \ldots, X_n; \theta_0)$ *is a test function for the hypothesis pair* $\{H_0 : \theta = \theta_0\}$ *against* $\{H_1 : \theta \neq \theta_0\}$ *with probability of type I error* α. *Then,*

$$R(X_1, \ldots, X_n) := \{\vartheta \in \Theta : \delta(X_1, \ldots, X_n; \vartheta) = 0\}$$

is a $(1 - \alpha)$*-confidence region for* θ.

Proof of Theorem 5.10 We first prove part (1). It suffices to show that the level of the test δ is α. But observe that

$$\mathbb{P}_{\theta_0}[\delta(X_1, \ldots, X_n) = 1] = 1 - \mathbb{P}_{\theta_0}[\delta(X_1, \ldots, X_n) = 0]$$

$$= 1 - \mathbb{P}_{\theta_0}[L(X_1, \ldots, X_n) \leq \theta_0 \leq U(X_1, \ldots, X_n)]$$

$$\leq 1 - \inf_{\theta \in \Theta} \mathbb{P}_{\theta}[L(X_1, \ldots, X_n) \leq \theta \leq U(X_1, \ldots, X_n)]$$

$$= 1 - (1 - \alpha)$$

$$= \alpha.$$

so that the test is indeed a level α test. This proves part (1). Now we turn to part (2). We need to show that $R(X_1, \ldots, X_n)$ is a $(1 - \alpha)$-confidence region. Let us calculate

$$\mathbb{P}_{\theta}[R(X_1, \ldots, X_n) \ni \theta] = \mathbb{P}_{\theta}[\delta(X_1, \ldots, X_n; \theta) = 0]$$

$$= 1 - \mathbb{P}_{\theta}[\delta(X_1, \ldots, X_n; \theta) = 1]$$

$$= 1 - \alpha$$

where the last line follows from our assumption that δ has probability of type I error α, for all simple nulls. This proves (2) and completes the proof. $\square$

▶ **Remark 5.11** When we follow the process described in part (2) of Theorem 5.10 to get a region R from a test function δ, we speak of *inverting a test*.

▶ **Remark 5.12** Notice that in part (2) we say that $R(X_1, \ldots, X_n)$ is a region and not an interval. The reason is that, depending on the exact form of δ and the model $f(x; \theta)$, the set $R(X_1, \ldots, X_n)$ may be a union of intervals, or perhaps even a more complicated set. For some forms of δ and some models $f(x; \theta)$, the region $R(X_1, \ldots, X_n)$ is indeed an interval. It is not hard to check that likelihood ratio tests and Wald tests for one-parameter exponential families do indeed yield a region $R(X_1, \ldots, X_n)$ that is an interval.

Example 5.13 (Mean of a Gaussian)

Compare the form of the test in Example (4.22, p. 116) with the form of the two-sided confidence interval in Exercise (59, p. 135) and conclude that the test and the interval are dual to each other.

Example 5.14 (Wald Tests and Intervals)

Compare the form of the approximate Wald test in the example following Theorem (4.26, p. 122) with the form of the two-sided approximate Wald interval in Exercise (63, p. 140).

Example 5.15 (Likelihood Ratio Tests and Intervals)

Compare the form of the approximate likelihood ratio test in the example following Theorem (4.23, p. 120) with the form of the two-sided approximate likelihood ratio interval discussed after the proof of Proposition (5.9, p. 140).

Note that in Theorem 5.10, we only considered two-sided intervals and tests. What about unilateral intervals and tests? For unilateral results, one direction is very easy: if $(-\infty, U]$ is a right-sided $(1 - \alpha)$-confidence interval for θ, then $\delta = \mathbf{1}\{U < \theta_0\}$ is a level α test for $\{H_0 : \theta \geq \theta_0\}$ vs $\{H_1 : \theta < \theta_0\}$ (and symmetrically for right-sided intervals).[3] So it's still easy to get unilateral hypothesis tests from unilateral confidence intervals. The opposite direction is more complicated. Getting a unilateral interval from a unilateral test depends on the form of the test function and on the form of the model under consideration.[4] Below we give a case where it's possible.

---

[3] The proof of this is analogous to the first part of Theorem 5.10.

[4] The problem is that, as we saw in Theorem 5.10, we have no guarantee in general that the region we get from inverting a test will be an interval, much less so a "one-sided" interval, unless there are further conditions.

Proposition 5.16 (One-Sided Intervals from Unilateral Tests) *Let $X_1, \ldots, X_n$ be an iid sample from a one-parameter exponential family with density (or frequency)*

$$f(x; \theta) = \exp\{\eta(\theta)T(x) - d(\theta) + S(x)\}, \qquad x \in \mathcal{X}, \ \theta \in \Theta \subseteq \mathbb{R}.$$

such that $\eta(\cdot)$ is strictly increasing and continuously differentiable, and Θ is open. Assume that $\tau = \sum_{i=1}^{n} T(X_i)$ is a continuous random variable, with distribution function $\mathbb{P}_\theta[\tau \leq t] = G(t; \theta)$.

1. Let $\delta(X_1, \ldots, X_n; \theta_0)$ be the UMP test of

$$\left\{ \begin{array}{l} H_0 : \theta \leq \theta_0 \\ H_1 : \theta > \theta_0 \end{array} \right\}$$

at level α, as defined in Theorem (4.16, p. 109). Then, the region

$$R(X_1, \ldots, X_n) = \{\vartheta \in \Theta : \delta(X_1, \ldots, X_n; \vartheta) = 0\}$$

is a $(1 - \alpha)$ left-sided interval of the form $[L(X_1, \ldots, X_n), +\infty)$.

2. Let $\delta(X_1, \ldots, X_n; \theta_0)$ be the UMP test of

$$\left\{ \begin{array}{l} H_0 : \theta \geq \theta_0 \\ H_1 : \theta < \theta_0 \end{array} \right\}$$

at level α, as defined in Theorem (4.16, p. 109). Then, the region

$$R(X_1, \ldots, X_n) = \{\vartheta \in \Theta : \delta(X_1, \ldots, X_n; \vartheta) = 0\}$$

is a $(1 - \alpha)$ right-sided interval of the form $(-\infty, U(X_1, \ldots, X_n)]$.

Proof We will prove only part (1), as (2) will then follow by symmetric arguments. The form of the test function $\delta(X_1, \ldots, X_n; \vartheta)$ is given by Theorem (4.16, p. 109) to be

$$\delta(X_1, \ldots, X_n; \theta_0) = \mathbf{1}\{\tau(X_1, \ldots, X_n) > q_{1-\alpha}(\theta_0)\} = \mathbf{1}\{\tau(X_1, \ldots, X_n) \geq q_{1-\alpha}(\theta_0)\}$$

where $q_{1-\alpha}(\theta_0)$ is the $(1 - \alpha)$-quantile of $G(t; \theta_0)$. It follows that

$$\begin{aligned} R(X_1, \ldots, X_n) &= \{\vartheta \in \Theta : \tau(X_1, \ldots, X_n) < q_{1-\alpha}(\vartheta)\} \\ &= \{\vartheta \in \Theta : G(\tau(X_1, \ldots, X_n); \vartheta) < G(q_{1-\alpha}(\vartheta); \vartheta)\} \\ &= \{\vartheta \in \Theta : G(\tau(X_1, \ldots, X_n); \vartheta) < 1 - \alpha\} \\ &= \{\vartheta \in \Theta : 1 - G(\tau(X_1, \ldots, X_n); \vartheta) > \alpha\} \end{aligned}$$

where the second equality follows since $G(t; \vartheta)$ is non-decreasing in t for all ϑ. If we can show that $G(t; \vartheta)$ is continuous with respect to ϑ, then the region R will necessarily be a union of intervals. If f we can also show that $1 - G(t; \vartheta) = \mathbb{P}_\vartheta[\tau(X_1, \ldots, X_n) > t]$ is increasing in ϑ for all t, then it will be clear that R will in fact be a single contiguous interval of the form $[L, +\infty)$, for some random variable L. But under our conditions, $\mathbb{P}_\vartheta[\tau(X_1, \ldots, X_n) > t]$ has indeed been proven to be differentiable and increasing in ϑ in the first part of the proof of Theorem (4.16, p. 109).[5] To complete the proof, we need to show that the confidence level of $R(X_1, \ldots, X_n) = [L(X_1, \ldots, X_n), +\infty)$ is indeed $1 - \alpha$. This follows easily by observing that for any $\vartheta \in \Theta$:

$$\mathbb{P}_\vartheta[L(X_1, \ldots, X_n) \le \vartheta] = \mathbb{P}_\vartheta[R(X_1, \ldots, X_n) \ni \vartheta] = \mathbb{P}_\vartheta[\delta(X_1, \ldots, X_n; \vartheta) = 0]$$

$$= \mathbb{P}_\vartheta[\tau(X_1, \ldots, X_n) \le q_{1-\alpha}(\vartheta)]$$

$$= G(q_{1-\alpha}(\vartheta); \vartheta)$$

$$= 1 - \alpha.$$

□

In non-technical terms, the theorem says that under some conditions, inverting a one-sided test in an exponential family will give a one-sided confidence interval. The details of exactly how this interval is constructed are not the most essential part here. The important thing is that we have found that the optimal one-sided tests can be used to yield confidence intervals. Since the tests are optimal, should the intervals not be optimal too? But what do we mean by an optimal confidence interval? We will consider these questions in the next paragraph.

5.4   Optimality in Interval Estimation

When discussing hypothesis tests, we saw that there are cases (depending on the hypothesis pair structure) where there was an optimal test function that one should use. It is therefore natural to wonder whether there are also cases in interval estimation, where there is an optimal confidence interval that one should use. How should one define optimal, though? It seems that any definition of optimality should satisfy the following two criteria:
1. Intuitively, optimal confidence intervals should be as "short" as possible on average, subject to being able to respect their confidence level: the shorter the interval, the more precise our localisation of our parameter.

---

[5]Recall that in that theorem we proved that the derivative of the mapping $\vartheta \mapsto \mathbb{E}_\vartheta[\delta(X_1, \ldots, X_n)] = \mathbb{P}_\vartheta[\tau \ge c]$ exists and is positive for all ϑ and all c.

2. Mathematically, we have seen that there exists a natural duality between confidence intervals and hypothesis tests. Therefore, any notion of optimality for confidence intervals should be dual to the notion of optimality for hypothesis tests. In other words: inverting an optimal hypothesis test should give us an optimal confidence interval.

Since we have seen that in general there can be no optimal test in bilateral hypothesis pairs, the second criterion rules out hopes of being able to obtain optimal two-sided confidence intervals. What about one-sided intervals, though? It turns out that the following definition of optimality for one-sided intervals satisfies both of the stated criteria:

Definition 5.17 (Uniformly Most Accurate One-Sided Intervals)
Let $[L(X_1,\ldots,X_n),+\infty)$ and $[M(X_1,\ldots,X_n),+\infty)$ be two left-sided $(1-\alpha)$ confidence intervals for θ. If for all $\theta \in \Theta$,

$$\mathbb{P}_\theta[\theta - L \geq \epsilon] \leq \mathbb{P}_\theta[\theta - M \geq \epsilon], \qquad \forall \epsilon > 0,$$

then $[L(X_1,\ldots,X_n),+\infty)$ is said to be more accurate than $[M(X_1,\ldots,X_n),+\infty)$ at confidence level $1-\alpha$. If $[L,+\infty)$ is more accurate than any other competing $(1-\alpha)$ left-sided interval, then it is called a uniformly most accurate (UMA) left-sided interval at confidence level $(1-\alpha)$.
Let $(-\infty,U(X_1,\ldots,X_n)]$ and $(-\infty,M(X_1,\ldots,X_n)]$ be two right-sided $(1-\alpha)$ confidence intervals for θ. If for all $\theta \in \Theta$,

$$\mathbb{P}_\theta[U - \theta \geq \epsilon] \leq \mathbb{P}_\theta[M - \theta \geq \epsilon], \qquad \forall \epsilon > 0,$$

then $(-\infty,U(X_1,\ldots,X_n)]$ is said to be more accurate than $(-\infty,M(X_1,\ldots,X_n)]$ at confidence level $1-\alpha$. If $(-\infty,U)$ is more accurate than any other competing $(1-\alpha)$ right-sided interval, then it is called a uniformly most accurate (UMA) right-sided interval at confidence level $(1-\alpha)$.

▶ **Remark 5.18 (On Interpreting the Optimality of Intervals)** Since one-sided intervals have infinite length, we cannot really make sense of what it means to have a "shortest" interval. Therefore, we define a one-sided interval to be most accurate if the bound it provides is less likely to be at a distance larger than $\epsilon > 0$ from the true parameter than any other competing interval, whatever the true parameter may be, and whatever $\epsilon > 0$ may be. Loosely speaking, the average tightness of a most accurate interval's bound is higher than the average tightness of any other interval's bound. Figure 5.1 provides a visual illustration of the concept.

Our definition can be seen to satisfy the requirement of intuitively being equivalent to "shortness" of the confidence intervals. The next proposition establishes that it also respects the duality with hypothesis tests (at least within the context of one-parameter exponential families) in the sense that the inversion of the uniformly most powerful hypothesis test yields the most accurate confidence interval.

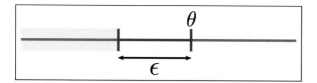

Fig. 5.1 Illustration of the definition of a most accurate *left-sided* interval. The idea is that, given $\varepsilon > 0$, the optimal interval's lower bound $L(X_1, \ldots, X_n)$ is less likely to fall in the shaded region than the lower bound of any other *left-sided* interval (in both cases subject to the constraint of having confidence level $1 - \alpha$)

Proposition 5.19 (UMP Tests $\Rightarrow$ UMA Intervals in Exponential Families)
Let $X_1, \ldots, X_n$ be an iid sample from a one-parameter exponential family with density (or frequency)

$$f(x; \theta) = \exp\{\eta(\theta)T(x) - d(\theta) + S(x)\}, \qquad x \in \mathcal{X}, \, \theta \in \Theta \subseteq \mathbb{R}.$$

such that $\eta(\cdot)$ is strictly increasing and continuously differentiable, and Θ is open. Assume that $\tau = \sum_{i=1}^{n} T(X_i)$ is a continuous random variable, with distribution function $\mathbb{P}_\theta[\tau \leq t] = G(t; \theta)$.
Given any $\theta_0 \in \Theta$, let $\delta(X_1, \ldots, X_n; \theta_0)$ be the UMP test of

$$\left\{ \begin{array}{l} H_0 : \theta \leq \theta_0 \\ H_1 : \theta > \theta_0 \end{array} \right\}$$

at level α. Then, the region

$$R(X_1, \ldots, X_n) = \{\vartheta \in \Theta : \delta(X_1, \ldots, X_n; \vartheta) = 0\}$$

is a uniformly most accurate $(1 - \alpha)$ left-sided confidence interval at confidence level $1 - \alpha$.

▶ **Remark 5.20** Of course, the symmetric version of this theorem holds true for right-sided intervals.

Proof From Proposition 5.16 (p. 144) we know that $R(X_1, \ldots, X_n)$ is a confidence interval of the form $[L(X_1, \ldots, X_n), +\infty)$, for some statistic L, whose confidence level is $1 - \alpha$. So $R(X_1, \ldots, X_n)$ is indeed a left-sided $(1 - \alpha)$ confidence interval. Therefore, it suffices to show that $[L, +\infty)$ is uniformly most accurate. To this aim, let $[M(X_1, \ldots, X_n), +\infty)$ be any other $1 - \alpha$ left-sided confidence interval. Define $\psi(X_1, \ldots, X_n; \theta) = \mathbf{1}\{M(X_1, \ldots, X_n) > \theta\}$ to be its dual test, which will have level α (to see this, follow the same steps as just above, replacing L by M). Given a $\theta_1 \in \Theta$ and an $\epsilon > 0$, define $\theta_0 = \theta_1 - \epsilon$ (so that $\theta_1 > \theta_0$). Since $\delta(X_1, \ldots, X_n; \theta_0)$

is UMP, we have:

$$\mathbb{P}_{\theta_1}[\delta(X_1, \ldots, X_n; \theta_0) = 1] \geq \mathbb{P}_{\theta_1}[\psi(X_1, \ldots, X_n; \theta_0) = 1]$$
$$\implies \mathbb{P}_{\theta_1}[\theta_0 < L(X_1, \ldots, X_n)] \geq \mathbb{P}_{\theta_1}[\theta_0 < M(X_1, \ldots, X_n)]$$
$$\implies \mathbb{P}_{\theta_1}[L(X_1, \ldots, X_n) \leq \theta_0] \leq \mathbb{P}_{\theta_1}[M(X_1, \ldots, X_n) \leq \theta_0]$$
$$\implies \mathbb{P}_{\theta_1}[\theta_0 \geq L(X_1, \ldots, X_n)] \leq \mathbb{P}_{\theta_1}[\theta_0 \geq M(X_1, \ldots, X_n)]$$
$$\implies \mathbb{P}_{\theta_1}[\theta_1 - \epsilon \geq L(X_1, \ldots, X_n)] \leq \mathbb{P}_{\theta_1}[\theta_1 - \epsilon \geq M(X_1, \ldots, X_n)]$$
$$\implies \mathbb{P}_{\theta_1}[\theta_1 - L \geq \epsilon] \leq \mathbb{P}_{\theta_1}[\theta_1 - M \geq \epsilon].$$

Since $\theta_1 \in \Theta$ and ϵ were arbitrary, we have established that $[L, +\infty)$ is more accurate than $[M, +\infty)$.   □

Exercise 65 Let $X_1, \ldots, X_n \overset{iid}{\sim} N(\mu, \sigma^2)$, where σ^2 is known. Find the expression for the UMA left-sided interval for μ at confidence level $1 - \alpha$.

Exercise 66 Let $X_1, \ldots, X_n \overset{iid}{\sim} Bern(p)$. Using the sufficient statistic $\tau_n(X_1, \ldots, X_n)$ for p, find the UMA left-sided interval for p at confidence level $1 - \alpha$, by inverting the test

$$H_0 : p \leq p_0 \qquad \text{vs} \qquad H_1 : p > p_0.$$

The endpoints of this interval are not as explicit as in the previous exercise. Unfortunately, one of the conditions of Proposition 5.19 is not satisfied (which one?). Thus, for most value of p, the coverage probability will only approximately be $1 - \alpha$.

Exercise 67 Show that the uniformly most accurate interval in Proposition 5.19 coincides with the interval constructed using the pivot $Y_n = F_{\tau_n}(\tau_n)$, as in Exercise 64, p. 141.

5.5   On Interpreting Confidence Intervals

It is very important to take care when interpreting the meaning of a confidence interval. Notice that

$$\inf_{\theta \in \Theta} \mathbb{P}_\theta \left[L(X_1, \ldots, X_n) \leq \theta \leq U(X_1, \ldots, X_n) \right] \geq 1 - \alpha$$

is an equivalent statement to

$$\inf_{\theta \in \Theta} \mathbb{P}_\theta \left\{ \theta \in \left[L(X_1, \ldots, X_n), U(X_1, \ldots, X_n) \right] \right\} \geq 1 - \alpha.$$

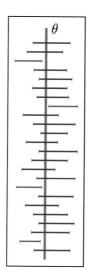

Fig. 5.2 Visualising the notion of a confidence interval at confidence level $1 - \alpha$. The *vertical line* represents the location of the fixed parameter value on the real axis. The parallel *black lines* represent realisations of the random interval $[L, U]$ for $r = 24$ different random samples from $f(x; \theta)$. We can see that most of them cover θ but some of them fail to do so. By the law of large numbers, we expect that as the number of replications $r \to \infty$, the proportion of intervals not covering θ will gradually converge to a number smaller than α

Though mathematically these statements are equivalent, the second way of writing the statement may lead to a misinterpretation of what an confidence interval means.

Specifically, it is the interval $[L, U]$ that is random and not the parameter θ. So saying that "the probability that the parameter fall inside the interval is at least $1-\alpha$" is wrong: the parameter is not going or falling anywhere, it is fixed. It is the interval that may change for different samples $X_1, .., X_n$, and may or may not cover the parameter (see Fig. 5.2). Therefore, one should say "the probability that the interval cover the parameter θ is at least $(1 - \alpha)$".

A different way of clarifying this is by noticing that:

$$\mathbb{P}_\theta\Big[L(X_1, \ldots, X_n) \leq \theta \leq U(X_1, \ldots, X_n)\Big] = \mathbb{P}_\theta\Big[\{L(X_1, \ldots, X_n) \leq \theta\}\} \cap \{U(X_1, \ldots, X_n) \geq \theta\}\Big]$$

where the right-hand side emphasises that the probability statement applies to the random confidence limits L and U, rather than to the deterministic parameter θ. To make sure that we avoid confusion, it is better to write $\mathbb{P}_\theta \{[L, U] \ni \theta\}$ instead of $\mathbb{P}_\theta \{\theta \in [L, U]\}$.

Concluding this section, we give two exercises which show how the notion of confidence intervals (and dual tests) admits a much more general interpretation when considering vectors of several parameters.

Exercise 68 Let $X_1, \ldots, X_n$ be random vectors in $\mathbb{R}^2$, defined as $X_i = (X_{i1}, X_{i2})^\top$, where $X_{11}, \ldots, X_{n1} \overset{iid}{\sim} N(\mu_1, \sigma^2)$, $X_{12}, \ldots, X_{n2} \overset{iid}{\sim} N(\mu_2, \sigma^2)$, and the $\{X_{i1}\}_{i=1}^n$ are independent of the $\{X_{i2}\}_{i=1}^n$. Assuming that σ is known, we wish to construct a confidence region for the parameter vector $\mu = (\mu_1, \mu_2)^\top$, that is, a random subset $C(X_1, \ldots, X_n)$ of $\mathbb{R}^2$ satisfying

$$\mathbb{P}_\mu\left[\mu \in C(X_1, \ldots, X_n)\right] \geq 1 - \alpha, \qquad \forall \mu \in \mathbb{R}^2$$

for a certain given confidence level $1 - \alpha$, $\alpha \in (0, 1)$.
1. Consider confidence regions for $\mu = (\mu_1, \mu_2)^\top$ of the form:

$$C_1(X_1, \ldots, X_n) = \left\{ \mu \in \mathbb{R}^2 : \bar{X}_1 - z_{1-\alpha'/2} \frac{\sigma}{\sqrt{n}} \leq \mu_1 \leq \bar{X}_1 + z_{1-\alpha'/2} \frac{\sigma}{\sqrt{n}}, \right.$$

$$\left. \bar{X}_2 - z_{1-\alpha'/2} \frac{\sigma}{\sqrt{n}} \leq \mu_2 \leq \bar{X}_2 + z_{1-\alpha'/2} \frac{\sigma}{\sqrt{n}} \right\}.$$

Find the value of α' for which $C_1(X_1, \ldots, X_n)$ is a confidence region with confidence level $1 - \alpha$.
2. Consider now confidence regions for μ of a different form, namely:

$$C_2(X_1, \ldots, X_n) = \left\{ \mu \in \mathbb{R}^2 : \frac{n}{\sigma^2}\left((\bar{X}_1 - \mu_1)^2 + (\bar{X}_2 - \mu_2)^2\right) \leq Q \right\}.$$

Find the value of Q for which $C_2(X_1, \ldots, X_n)$ is a confidence region for μ with confidence level $1 - \alpha$.
3. Let $\bar{X}_1 = -0.7$, $\bar{X}_2 = 0.6$, $n = 9$, $\sigma^2 = 1$. Draw the regions C_1 and C_2 at confidence level 95% on the plane $\mathbb{R}^2$. Find the ratio of the areas of the two regions. Which is preferable?

Exercise 69 In the same notation and under the same assumptions as in the previous exercise, construct two test functions to test $H_0 : \mu = 0$ vs $H_1 : \mu \neq 0$ at level $\alpha \in (0, 1)$, by inverting the previous regions.

Appendix

A.1   Probability Factsheet

This section provides a snapshot of some of the main probabilistic concepts and properties that are made use of in the text. For a more detailed coverage, the reader is referred to Knight [14], Chaps. 1–3.

Events

A random experiment is a process whose outcome is uncertain. The possible outcomes, and combinations thereof, are described in the language of set theory. In principle, any statement that makes reference to the outcome of a random experiment should be expressible via this language. In detail:
- A possible outcome ω of a random experiment is called an elementary event.
- The set of all possible outcomes, say Ω is assumed non-empty, $\Omega \neq \emptyset$.
- An event is a subset $F \subset \Omega$ of Ω. An event F "is realised" (or "occurs") whenever the outcome of the experiment is an element of F.
- The union of two events F_1 and F_2, written $F_1 \cup F_2$ occurs if and only if either of F_1 or F_2 occurs. Equivalently, $\omega \in F_1 \cup F_2$ if and only if $\omega \in F_1$ or $\omega \in F_2$,

$$F_1 \cup F_2 = \{\omega \in \Omega : \omega \in F_1 \text{ or } \omega \in F_2\}$$

- The intersection of two events F_1 and F_2, written $F_1 \cap F_2$ occurs if and only if both F_1 and F_2 occur. Equivalently, $\omega \in F_1 \cap F_2$ if and only if $\omega \in F_1$ and $\omega \in F_2$,

$$F_1 \cap F_2 = \{\omega \in \Omega : \omega \in F_1 \text{ and } \omega \in F_2\}$$

- Unions and intersections of several events, $F_1 \cup \ldots \cup F_n$ and $F_1 \cap \ldots \cap F_n$ are defined iteratively from the definition for unions and intersections of pairs.

© Springer International Publishing Switzerland 2016
V.M. Panaretos, *Statistics for Mathematicians*, Compact Textbooks in Mathematics,
DOI 10.1007/978-3-319-28341-8

- The complement of an event F, denoted F^c, contains all the elements of Ω that are not contained in F,

$$F^c = \{\omega \in \Omega : \omega \notin F\}.$$

- Two events F_1 and F_2 are called disjoint if they contain no common elements, that is $F_1 \cap F_2 = \emptyset$.
- A partition $\{F_n\}_{n \geq 1}$ of Ω is a collection of events such that $F_i \cap F_j = \emptyset$ for all $i \neq j$, and $\cup_{n \geq 1} F_n = \Omega$.
- The difference of two events F_1 and F_2 is defined as $F_1 \backslash F_2 = F_1 \cap F_2^c$. It contains all the elements of F_1 that are not contained in F_2. Notice that the difference is not symmetric: $F_1 \backslash F_2 \neq F_2 \backslash F_1$.
- It can be checked that the following properties hold true
  - (i) $(F_1 \cup F_2) \cup F_3 = F_1 \cup (F_2 \cup F_3) = F_1 \cup F_2 \cup F_3$
  - (ii) $(F_1 \cap F_2) \cap F_3 = F_1 \cap (F_2 \cap F_3) = F_1 \cap F_2 \cap F_3$
  - (iii) $F_1 \cap (F_2 \cup F_3) = (F_1 \cap F_2) \cup (F_1 \cap F_3)$
  - (iv) $F_1 \cup (F_2 \cap F_3) = (F_1 \cup F_2) \cap (F_1 \cup F_3)$
  - (v) $(F_1 \cup F_2)^c = F_1^c \cap F_2^c$ and $(F_1 \cap F_2)^c = F_1^c \cup F_2^c$

Probability Axioms

A probability measure $\mathbb{P}$ is a real function defined over the events of Ω, assigning a probability to any event. This can be interpreted as a measure of how certain we are that the event will occur. It is postulated to satisfy the following properties:
1. $\mathbb{P}(F) \geq 0$, for all events F.
2. $\mathbb{P}(\Omega) = 1$.
3. If $\{F_n\}_{n \geq 1}$ are disjoint events, and $F = \cup_{n \geq 1} F_n$ is an event given by their union, then

$$\mathbb{P}(F) = \sum_{n \geq 1} \mathbb{P}(F_n).$$

The following properties are immediate consequences of the probability axioms:
- $\mathbb{P}(F^c) = 1 - \mathbb{P}(F)$.
- $\mathbb{P}(F_1 \cap F_2) \leq \min\{\mathbb{P}(F_1), \mathbb{P}(F_2)\}$.
- $\mathbb{P}(F_1 \cup F_2) = \mathbb{P}(F_1) + \mathbb{P}(F_2) - \mathbb{P}(F_1 \cap F_2)$.
- Continuity from below: let $\{F_n\}_{n \geq 1}$ be nested events, such that $F_j \subseteq F_{j+1}$ for all j, and let F be an event given by $F = \cup_{n \geq 1} F_n$. Then $\mathbb{P}(F_n) \overset{n \to \infty}{\longrightarrow} \mathbb{P}(F)$.
- Continuity from above: let $\{F_n\}_{n \geq 1}$ be nested events, such that $F_j \supseteq F_{j+1}$ for all j, and let F be an event given by $F = \cap_{n \geq 1} F_n$. Then $\mathbb{P}(F_n) \overset{n \to \infty}{\longrightarrow} \mathbb{P}(F)$.
- If $\Omega = \{\omega_1, \ldots, \omega_K\}$, $K < \infty$, is a finite set, then for any event $F \subseteq \Omega$, we have $\mathbb{P}(F) = \sum_{j:\omega_j \in F} \mathbb{P}(\omega_j)$.

Conditional Probability and Independence

Suppose we do not know the precise outcome $\omega \in \Omega$ that has occurred, but we are told that $\omega \in F_2$ for some event F_2. If we are asked to now calculate the probability that $\omega \in F_1$ also, for some other event F_1, then we need to calculate the conditional probability of F_1 given F_2.

- For any pair of events F_1, F_2 such that $\mathbb{P}(F_2) > 0$, we define the conditional probability of F_1 given F_2 to be

$$\mathbb{P}(F_1|F_2) = \frac{\mathbb{P}(F_1 \cap F_2)}{\mathbb{P}(F_2)}.$$

- Let G be an event and $\{F_n\}_{n \geq 1}$ be a partition of Ω such that $\mathbb{P}(F_n) > 0$ for all n. We then have:
  - Law of total probability:

$$\mathbb{P}(G) = \sum_{n=1}^{\infty} \mathbb{P}(G|F_n)\mathbb{P}(F_n)$$

  - Bayes' theorem:

$$\mathbb{P}(F_j|G) = \frac{\mathbb{P}(F_j \cap G)}{\mathbb{P}(G)} = \frac{\mathbb{P}(G|F_j)\mathbb{P}(F_j)}{\sum_{n=1}^{\infty} \mathbb{P}(G|F_n)\mathbb{P}(F_n)}$$

- The events $\{G_n\}_{n \geq 1}$ are called independent if and only if for any finite sub-collection $\{G_{i_1}, \ldots, G_{i_K}\}$, $K < \infty$, we have:

$$\mathbb{P}(G_{i_1} \cap \cdots \cap G_{i_K}) = \mathbb{P}(G_{i_1}) \times \mathbb{P}(G_{i_2}) \times \ldots \times \mathbb{P}(G_{i_K})$$

Random Variables and Distribution Functions

Random variables are, simply stated, numerical summaries of the outcome of a random experiment. Since the result is random, such numerical summaries are random, too. They allow us to not worry too much about the precise structure of the outcome $\omega \in \Omega$, but concentrate on a numerical summary instead. If that numerical summary is all we really care about, we can concentrate on the range of a random variable X, rather than consider Ω itself.

- A random variable is a real function $X : \Omega \to \mathbb{R}$.
- We write $\{a \leq X \leq b\}$ to denote the event

$$\{\omega \in \Omega : a \leq X(\omega) \leq b\}.$$

More generally, if $A \subset \mathbb{R}$ is a more general subset, we write $\{X \in A\}$ to denote the event

$$\{\omega \in \Omega : X(\omega) \in A\}.$$

- If we have a probability measure defined on the events of Ω, then X induces a new probability measure on subsets of the real line. This is described by the distribution function (or cumulative distribution function) $F_X : \mathbb{R} \to [0, 1]$ of a random variable X (or the law of X). This is defined as

$$F_X(x) = \mathbb{P}(X \leq x).$$

- By its definition, a distribution function satisfies the following properties:
    (i) $x \leq y \Rightarrow F_X(x) \leq F_X(y)$
    (ii) $\lim_{x \to \infty} F_X(x) = 1$, $\lim_{x \to -\infty} F_X(x) = 0$
    (iii) $\lim_{y \downarrow x} F_X(y) = F_X(x)$, that is, F_X is right-continuous.
    (iv) $\lim_{y \uparrow x} F_X(y)$ exists, that is, F_X is left-limited.
    (v) $\mathbb{P}(a < X \leq b) = F_X(b) - F_X(a)$.
    (vi) $\mathbb{P}(X > a) = 1 - F(a)$.
    (vii) Let $D_X := \{x \in \mathbb{R} : F_X(x) - \lim_{y \uparrow x} F_X(y) > 0\}$ be the set of points where F_X is not continuous.
        - D_X is a countable set (Lemma A.11, p. 169).
        - If $\mathbb{P}(\{X \in D_F\}) = 1$, then X is called a *discrete* random variable (equivalently, X has a finite or countable range, with probability 1).
        - If $D_X = \emptyset$, then X is called a *continuous* random variable (the distribution function F_X is continuous).
        - It may very well happen that a random variable may be neither discrete nor continuous.

Probability Density and Probability Mass Functions

- The probability mass function (or frequency function) $f_X : \mathbb{R} \to [0, 1]$ of a discrete random variable X is defined as

$$f_X(x) = \mathbb{P}(X = x).$$

By its definition, a probability mass function satisfies
(i) $\mathbb{P}(X \in A) = \sum_{t \in A \cap \mathcal{X}} f_X(t)$, for $A \subseteq \mathbb{R}$ and $\mathcal{X} = \{x \in \mathbb{R} : f_X(x) > 0\}$.
(ii) $F_X(x) = \sum_{t \in (-\infty, x] \cap \mathcal{X}} f_X(t)$, for all $x \in \mathbb{R}$ and $\mathcal{X} = \{x \in \mathbb{R} : f_X(x) > 0\}$.
(iii) An immediate corollary is that $F_X(x)$ is piecewise constant with jumps at the points in $\mathcal{X} = \{x \in \mathbb{R} : f_X(x) > 0\}$.

- A continuous random variable X has probability density function $f_X : \mathbb{R} \rightarrow [0, +\infty)$ if

$$F_X(b) - F_X(a) = \int_a^b f_X(t)dt.$$

for all real numbers $a < b$. By its definition, a probability density satisfies
(i) $F_X(x) = \int_{-\infty}^x f_X(t)dx$
(ii) $f_X(x) = F_X'(x)$, whenever f_X is continuous at x.
(iii) Note that $f_X(x) \neq \mathbb{P}(X = x) = 0$. In fact, it can be $f(x) > 1$ for some x. It can even happen that f is unbounded.

Random Vectors and Joint Distributions

A random vector $\mathbf{X} = (X_1, \ldots, X_d)^\top$ is a finite collection of random variables, arranged as the coordinates of a vector. The point is that we may want to make probabilistic statements on the joint behaviour of all these random variables. In this case, we need to define their joint distribution, and respective joint density (or joint frequency).

- The joint distribution function of a random vector $\mathbf{X} = (X_1, \ldots, X_d)^\top$ is defined as:

$$F_{\mathbf{X}}(x_1, \ldots, x_d) = \mathbb{P}(X_1 \leq x_1, \ldots, X_d \leq x_d).$$

- Correspondingly, one defines the
  - joint frequency function, if the $\{X_i\}_{i=1}^d$ are all discrete,

$$f_{\mathbf{X}}(x_1, \ldots, x_d) = \mathbb{P}(X_1 = x_1, \ldots, X_d = x_d).$$

  - the joint density function, if there exists $f_{\mathbf{X}} : \mathbb{R}^d \rightarrow [0, +\infty)$ such that:

$$F_{\mathbf{X}}(x_1, \ldots, x_d) = \int_{-\infty}^{x_1} \cdots \int_{-\infty}^{x_d} f_{\mathbf{X}}(u_1, \ldots, u_d)du_1 \ldots du_d$$

In this case, when $f_{\mathbf{X}}$ is continuous at the point $\mathbf{x}$,

$$f_{\mathbf{X}}(x_1, \ldots, x_d) = \frac{\partial^d}{\partial x_1 \ldots \partial x_d} F_{\mathbf{X}}(x_1, \ldots, x_d)$$

Marginal Distributions

Given the joint distribution of the random vector $\mathbf{X} = (X_1, \ldots, X_d)^\top$, we can always isolate the distribution of a single coordinate, say X_i.

- In the discrete case, the marginal frequency function of X_i is given by $f_{X_i} : \mathbb{R} \to [0, +\infty)$:

$$f_{X_i}(x_i) = \mathbb{P}(X_i = x_i) = \sum_{x_1} \cdots \sum_{x_{i-1}} \sum_{x_{i+1}} \cdots \sum_{x_d} f_X(x_1, \ldots, x_{i-1}, x_i, x_{i+1}, \ldots, x_d)$$

- In the continuous case, the marginal density function of X_i is given by $f_{X_i} : \mathbb{R} \to [0, +\infty)$:

$$f_{X_i}(x_i) = \int_{-\infty}^{\infty} \cdots \int_{-\infty}^{\infty} f_X(y_1, \ldots, y_{i-1}, x_i, y_{i+1}, \ldots, y_d) dy_1 \ldots dy_{i-1} dy_{i+1} dy_d.$$

- More generally, we can define the joint frequency/density of a random vector formed by a subset of the coordinates of $\mathbf{X} = (X_1, \ldots, X_d)^\top$, say the first k (with $k < d$), $(X_1, \ldots, X_k)^\top$, via
  - Discrete case: $f_{X_1,\ldots,X_k}(x_1, \ldots, x_k) = \sum_{x_{k+1}} \cdots \sum_{x_d} f_X(x_1, \ldots, x_k, x_{k+1}, \ldots, x_d)$.
  - Continuous case
    $f_{X_1,\ldots,X_k}(x_1, \ldots, x_k) = \int_{-\infty}^{+\infty} \cdots \int_{-\infty}^{+\infty} f_X(x_1, \ldots, x_k, x_{k+1}, \ldots, x_d) dx_{k+1} \ldots dx_d.$
- In other words, in order to find a marginal density/frequency of a subset of random variables, we need to integrate/sum out the remaining variables from the overall joint density/frequency.
- It is important to note that the marginals do not uniquely determine the joint distribution.

Conditional Distributions

Similarly to the notion of conditional probability, we may wish to make probabilistic statements about the potential outcomes of one random variable, if we already know the outcome of another. For this we need the notion of conditional density and conditional frequency functions. If $(X_1, \ldots, X_d)$ is a continuous/discrete random vector, we define the conditional probability density/frequency function of $(X_1, \ldots, X_k)$ given $\{X_{k+1} = x_{k+1}, \ldots, X_d = x_d\}$ as

$$f_{X_1,\ldots,X_k|X_{k+1},\ldots,X_d}(x_1, \ldots, x_k | x_{k+1}, \ldots, x_d) = \frac{f_{X_1,\ldots,X_d}(x_1, \ldots, x_k, x_{k+1}, \ldots, x_d)}{f_{X_{k+1},\ldots,X_d}(x_{k+1}, \ldots, x_d)}$$

provided that $f_{X_{k+1},\ldots,X_d}(x_{k+1}, \ldots, x_d) > 0$. The corresponding distribution functions are:
- In the discrete case:

$$F_{X_1,\ldots,X_k|X_{k+1},\ldots,X_d}(x_1, \ldots, x_k | x_{k+1}, \ldots, x_d)$$
$$= \sum_{u_1 \leq x_1} \cdots \sum_{u_k \leq x_k} f_{X_1,\ldots,X_k|X_{k+1},\ldots,X_d}(u_1, \ldots, u_k | x_{k+1}, \ldots, x_d).$$

• In the continuous case:

$$F_{X_1,\dots,X_k|X_{k+1},\dots,X_d}(x_1,\dots,x_k|x_{k+1},\dots,x_d)$$

$$= \int_{-\infty}^{x_1} \cdots \int_{-\infty}^{x_k} f_{X_1,\dots,X_k|X_{k+1},\dots,X_d}(u_1,\dots,u_k|x_{k+1},\dots,x_d)du_1\dots du_k.$$

Independent Random Variables

The random variables $X_1,\dots,X_d$ are called independent if and only if, for all $x_1,\dots,x_d \in \mathbb{R}$

$$F_{X_1,\dots,X_d}(x_1,\dots,x_d) = F_{X_1}(x_1) \times \dots \times F_{X_d}(x_d).$$

Equivalently, $X_1,\dots,X_d$ are independent if and only if, for all $x_1,\dots,x_d \in \mathbb{R}$

$$f_{X_1,\dots,X_d}(x_1,\dots,x_d) = f_{X_1}(x_1) \times \dots \times f_{X_d}(x_d).$$

Note that when random variables are independent, conditional distributions reduce to the corresponding marginal distributions. Intuitively, knowing the value of one of the random variables gives us no information about the distribution of the rest.

Expectation, Variance, Covariance

The expectation (or expected value) of a random variable X formalises the notion of the "average" value taken by that random variable (in a sense, the typical value, what we expect). It is defined as follows.
– For continuous variables:

$$\mathbb{E}[X] = \int_{-\infty}^{+\infty} x\, f_X(x)dx.$$

– For discrete variables:

$$\mathbb{E}[X] = \sum_{x \in \mathcal{X}} x\, f_X(x), \qquad \mathcal{X} = \{x \in \mathbb{R} : f_X(x) > 0\}.$$

The expectation satisfies the following properties:
• Linearity: $\mathbb{E}[X_1 + \alpha X_2] = \mathbb{E}[X_1] + \alpha\mathbb{E}[X_2]$.
• $\mathbb{E}[h(x)] = \sum_{x \in \mathcal{X}} h(x)f_X(x)$ (discrete case)
  or
  $\mathbb{E}[h(x)] = \int_{-\infty}^{+\infty} h(x)f(x)dx$ (continuous case).

The variance of a random variable X expresses how disperse the realisations of X are around its expectation.

$$\text{Var}(X) = \mathbb{E}\left[(X - \mathbb{E}(X))^2\right] \qquad (\text{if } \mathbb{E}[X^2] < \infty).$$

Furthermore, the covariance of a random variable X_1 with another random variable X_2 expresses the degree of linear dependency between the two.

$$\text{Cov}(X_1, X_2) = \mathbb{E}\left[(X_1 - \mathbb{E}(X_1))(X_2 - \mathbb{E}(X_2))\right] \qquad (\text{if } \mathbb{E}[X_i^2] < \infty).$$

The correlation between X_1 and X_2 is defined as

$$\text{Corr}(X_1, X_2) = \frac{\text{Cov}(X_1, X_2)}{\sqrt{\text{Var}(X_1)\,\text{Var}(X_2)}}.$$

It also expresses the degree of linear dependency. Its advantage is that it is invariant to changes of units of measurement, and moreover can be understood in absolute terms (it ranges in $[-1, 1]$), as a result of the correlation inequality (itself a consequence of the Cauchy–Schwarz inequality):

$$|\text{Corr}(X_1, X_2)| \le \sqrt{\text{Var}(X_1)\,\text{Var}(X_2)}.$$

Some useful formulae relating expectations, variance, and covariances are:
- $\text{Var}(X) = \mathbb{E}[X^2] - (\mathbb{E}[X])^2 = \text{Cov}(X, X)$
- $\text{Var}(aX + b) = a^2\,\text{Var}(X)$
- $\text{Var}(\sum_i X_i) = \sum_i \text{Var}(X_i) + \sum_{i \ne j} \text{Cov}(X_i, X_j)$
- $\text{Cov}(X_1, X_2) = \mathbb{E}[X_1 X_2] - \mathbb{E}[X_1]\mathbb{E}[X_2]$
- $\text{Cov}(aX_1 + bX_2, Y) = a\text{Cov}(X_1, Y) + b\text{Cov}(X_2, Y)$
- if $\mathbb{E}[X_1^2] + \mathbb{E}[X_2^2] < \infty$, then the following are equivalent:
    - (i) $\mathbb{E}[X_1 X_2] = \mathbb{E}[X_1]\mathbb{E}[X_2]$
    - (ii) $\text{Cov}(X_1, X_2) = 0$
    - (iii) $\text{Var}(X_1 \pm X_2) = \text{Var}(X_1) + \text{Var}(X_2)$

    Independence will imply these three last properties, but none of these properties imply independence.

A.2   Taylor's Formula and the Inverse Function Theorem

The following two classic analysis results will often be used. See Rudin [21] (Chaps. 5 and 9) for their proofs.[1]

---

[1] An elementary proof of the one-dimensional form of the inverse function theorem (which will be all that will be needed for this text as stated below) can also be found in Corwin and Szczarba [5], Chap. 9.

Theorem A.1 (Taylor's Formula with Lagrange Remainder) *Let $h(x) : \mathbb{R} \to \mathbb{R}$ be k-times continuously differentiable on the closed interval I with endpoints x and y, for some $k \geq 0$. If $f^{(k+1)}$ exists on the interior of I, then there exists $t \in (0, 1)$ such that*

$$h(x) = h(y) + h'(y)(x - y) + \frac{h''(y)}{2!}(x - y)^2 + \cdots + \frac{h^{(k)}(y)}{k!}(x - y)^k$$
$$+ \frac{h^{(k+1)}(\xi)}{(k + 1)!}(x - y)^{k+1}$$

for $\xi = tx + (1 - t)y$.

Theorem A.2 (Inverse Function Theorem) *Let $h(x) : \mathbb{R} \to \mathbb{R}$ be continuously differentiable, with a non-zero derivative at a point $x_o \in \mathbb{R}$. Then, there exists an $\varepsilon > 0$ such h^{-1} continuously differentiable on $(h(x_0) - \varepsilon, h(x_0) + \varepsilon)$, and in fact $(h^{-1})'(y) = [h'(h^{-1}(y))]^{-1}$ for $|y - h(x_0)| < \varepsilon$.*

A.3 Two Concentration Inequalities

Lemma A.3 (Markov's Inequality) *Let X be a non-negative random variable. Then, given any $\epsilon > 0$,*

$$\mathbb{P}[X \geq \epsilon] \leq \frac{\mathbb{E}[X]}{\epsilon}.$$

Proof Notice that $0 \leq \epsilon \mathbf{1}\{X \geq \epsilon\} \leq X$. Therefore, $\mathbb{E}[\epsilon \mathbf{1}\{X \geq \epsilon\}] \leq \mathbb{E}[X]$. But

$$\mathbb{E}[\epsilon \mathbf{1}\{X \geq \epsilon\}] = \epsilon \mathbb{E}[\mathbf{1}\{X \geq \epsilon\}] = \epsilon \left(1 \cdot \mathbb{P}[X \geq \epsilon] + 0 \cdot \mathbb{P}[X < \epsilon]\right) = \epsilon \mathbb{P}[X \geq \epsilon].$$

Combining our findings yields the result. □

Lemma A.4 (Chebyshev's Inequality) *Let X be a random variable with finite mean $\mathbb{E}[X] < \infty$. Then, given any $\epsilon > 0$,*

$$\mathbb{P}\left[|X - \mathbb{E}[X]| \geq \epsilon\right] \leq \frac{\text{Var}[X]}{\epsilon^2}.$$

Proof Define $Y = (X - \mathbb{E}[X])^2$ and apply Markov's inequality to Y. □

A.4   Monotonicity and Covariance

Lemma A.5 (Covariance of X and $g(X)$) *Let X be a real random variable with $\mathbb{E}[X^2] < \infty$. Let $g : \mathbb{R} \to \mathbb{R}$ be a non-decreasing function such that $\mathbb{E}[g^2(X)] < \infty$. Then,*

$$\mathrm{Cov}[X, g(X)] \geq 0.$$

Proof By definition of covariance:

$$\mathrm{Cov}[X, g(X)] = \mathbb{E}\Big\{\big(X - \mu\big)\big(g(X) - \mathbb{E}[g(X)]\big)\Big\}$$

$$= \mathbb{E}\Big\{\big(X - \mu\big)\big(g(X) - g(\mu) + g(\mu) - \mathbb{E}[g(X)]\big)\Big\}$$

$$= \mathbb{E}\Big\{\big(X - \mu\big)\big(g(X) - g(\mu)\big)\Big\} + \underbrace{\mathbb{E}\Big\{\big(X - \mu\big)\big(g(\mu) - \mathbb{E}[g(X)]\big)\Big\}}_{=0}$$

Now g is non-decreasing so if $X \geq \mu$, then $g(X) \geq g(\mu)$. If $X \leq \mu$, on the other hand, then $g(X) \leq g(\mu)$ also. Therefore

$$(X - \mu)(g(X) - g(\mu)) \geq 0$$

and the result follows.   □

A.5   Quantiles

Recall that, for a random variable X taking values in $\mathcal{X}$, we define its distribution function to be:

$$F_X : \mathbb{R} \to [0, 1],$$

$$F_X(x) = \mathbb{P}[X \leq x], \qquad x \in \mathbb{R}.$$

Simply put, the distribution function is the answer to the following question: given a real number $x \in \mathbb{R}$, what is the probability $\mathbb{P}[X \leq x]$ that X fall at or below x? We could also ask the opposite question:

Given a probability $\alpha \in (0, 1)$, is there a real number x such that $\mathbb{P}[X \leq x] = \alpha$?
                                                                                                (A.1)

The motivates the definition of the so-called *quantile function*.

Definition A.6 (Quantile Function and Quantiles)

Let X be a random variable and F_X be its distribution function. We define the quantile function of X to be the function

$$F_X^- : (0,1) \to \mathbb{R}$$

$$F_X^-(\alpha) = \inf\{t \in \mathbb{R} : F_X(t) \geq \alpha\}.$$

Given an $\alpha \in (0,1)$, we call the real number

$$q_\alpha = F_X^-(\alpha)$$

the α-quantile of X (or, equivalently, of F_X).

Recall that F_X is always non-decreasing, by its definition. Hence, there are two possibilities:

(A) F_X is in fact *strictly increasing*.[2] Then F_X is also invertible, and we have

$$F_X^-(\alpha) = F_X^{-1}(\alpha), \qquad \forall \alpha \in (0,1).$$

In this case, our question (A.1) has a unique answer, and the interpretation is very simple.

(B) F_X is non-decreasing, but not strictly increasing.[3] Then there are two things that may happen:

   (B1) There may be multiple real numbers x that satisfy $F_X(x) = \alpha$ (for example, take $\alpha = 1 - p$ and take X to be a Bern(p) random variable; then any $x \in (0,1)$ satisfies that $F_X(x) = 1 - p = \alpha$). In this case, $F_X^{-1}(\alpha)$ is a set, not a single real number,

$$F_X^{-1}(\alpha) = \{x \in \mathbb{R} : F_X(x) = \alpha\}.$$

So, which of these numbers should we pick as the answer to our question (A.1)? The most mathematically appropriate choice turns out to be the infimum of this set.[4] Since F_X is right-continuous (being a probability distribution function) the infimum of this set equals $F_X^-(\alpha)$.

---

[2] This is the case if X is continuous with a density that satisfies $f_X(x) > 0 \ \forall x \in \mathbb{R}$.

[3] For regular models, this happens if X is discrete (so F_X is a step-function) or when X is continuous but there exists at least one open interval I such that $f_X(x) = 0, \ \forall x \in I$.

[4] This is due to the fact that, with this definition, we have $F(X) \geq \alpha \iff X \geq F^{-1}(\alpha)$, which is very useful when generating random variables with a prescribed distribution, see Exercise 11 (p. 22).

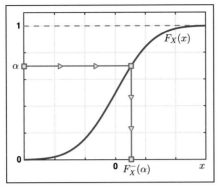

(a) Quantile in Scenario (A).

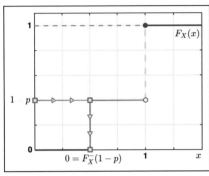

(b) Quantile in Scenario (B1).

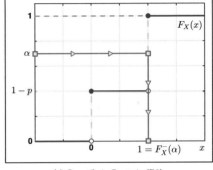

(c) Quantile in Scenario (B2).

Fig. A.1 Evaluation of the quantile function for scenario (A), (B1) and (B2) above. Intuitively, in order to find q_α, we follow the *red arrows*. (**a**) Quantile in Scenario (A). (**b**) Quantile in Scenario (B1). (**c**) Quantile in Scenario (B2)

(B2) There may be no real number x such that $F_X(x) = \alpha$ (for example, take some $\alpha \in (1 - p, 1)$ and take X to be a Bern(p) random variable). In this case, our question (A.1) has no answer. So, instead we have to settle for the first time that $F_X(x)$ "jumps" above α, which is again given by $F_X^-(x)$.

If all of this sounds complicated, Fig. A.1 gives an intuitive illustration that should clarify things.

Exercise 70 Let $X \sim \text{Exp}(\lambda)$ where $\lambda > 0$. Show that the α–quantile of X is given by

$$q_\alpha = F_X^-(\alpha) = -\log(1 - \alpha)/\lambda,$$

for $0 < \alpha < 1$.

Exercise 71 (Quantiles Determine Distributions) Let X and Y be random variables with respective distribution functions F_X and F_Y. Suppose that $F_X^-(\alpha) = F_Y^-(\alpha)$ for all $\alpha \in (0,1)$. Prove that $F_X = F_Y$.

A.6   Moment Generating Functions

The moment generating function (MGF) is a useful tool in probability theory that can often help us to prove independence of random variables or to determine their moments (hence the word moment generating).

> **Definition A.7 (Moment Generating Function)**
> Let X be a random variable taking values in $\mathbb{R}$. The MGF of X is defined as
>
> $$M_X(t) : \mathbb{R} \to \mathbb{R} \cup \{\infty\}$$
>
> $$M_X(t) = \mathbb{E}\left[e^{tX}\right], \qquad t \in \mathbb{R}.$$

Notice that $M_X(0) = 1$ always, so there exists at least one $t \in \mathbb{R}$ for which $M_X(t) < \infty$. When $M_X(t)$ is finite on an open neighbourhood of zero, then all the moments of X are defined, and can be determined by evaluating derivatives of M_X at zero.

> **Proposition A.8 (Moments via the MGF)** *Let X be a random variable taking values in $\mathbb{R}$, and let I be an open interval such that $M_X(t) < \infty$ for all $t \in I$. It holds that*
> 1. *$\mathbb{E}[|X|^k e^{tX}] < \infty$ for all $k \in \mathbb{N}$ and all $t \in I$.*
> 2. *For all $t \in I$, the function M_X is k times differentiable, for all $k \in \mathbb{N}$ (hence infinitely differentiable on I).*
> 3. *For all $k \in \mathbb{N}$ and all $t \in I$, $\mathbb{E}[X^k e^{tX}] = \frac{d^k M_X}{dt^k}(t)$.*
> 4. *If $\{0\} \subset I$, then $\mathbb{E}[|X|^k] < \infty$ and $\mathbb{E}[X^k] = \frac{d^k M_X}{dt^k}(0)$, for all $k \in \mathbb{N}$.*

Proof We start with part 1. Fix $t_0 \in I$ and $k \in \mathbb{N}$. Since I is open, there exists a $\delta > 0$ such that $[t_0 - \delta, t_0 + \delta] \subset I$. Since the exponential function is increasing, we have

$$|X|^k e^{t_0 X} = X^k e^{t_0 X} \mathbf{1}\{X \geq 0\} + (-X)^k e^{t_0 X} \mathbf{1}\{X < 0\}$$

$$= e^{(t_0+\delta)X} u_{k,\delta}(X)\mathbf{1}\{X \geq 0\} + e^{(t_0-\delta)X} u_{k,\delta}(-X)\mathbf{1}\{X < 0\},$$

where $u_{k,\delta} : [0,\infty) \to [0,\infty)$ is given by

$$u_{k,\delta}(x) = x^k \exp(-\delta x), \qquad k \geq 0, \quad \delta > 0, \quad x \geq 0.$$

It's not hard to see that $C_{k,\delta} = \sup_{x \geq 0} u_{k,\delta}(x) < \infty$, since the exponential will decay faster than any polynomial. Specifically,

$$u'_{k,\delta}(x) = x^{k-1}e^{-\delta x}(k-\delta x) \begin{cases} > 0 & x < \frac{k}{\delta} \\ < 0 & x > \frac{k}{\delta}, \end{cases}$$

so that $u_{k,\delta}$ attains its maximum at $x = k/\delta$. We conclude that

$$\mathbb{E}|X|^k e^{t_0 X} \leq C_{k,\delta}\mathbb{E}e^{(t_0+\delta)X}\mathbb{1}\{X \geq 0\} + C_{k,\delta}\mathbb{E}e^{(t_0-\delta)X}\mathbb{1}\{X < 0\}$$

$$\leq C_{k,\delta}M_X(t_0+\delta) + C_{k,\delta}M_X(t_0+\delta) < \infty.$$

Since the choice of t_0 was arbitrary, we have proven part 1.

In order to prove parts 2 and 3, we proceed recursively. Both parts are trivially valid when $k = 0$. We will now show that if 2 and 3 are valid for $k - 1$ (for all $t \in I$), then they must be valid for k, whenever $k \geq 1$.

Fix $t_0 \in I$. We need to show that

$$\lim_{t \to t_0} \frac{\mathbb{E}X^{k-1}e^{tX} - \mathbb{E}X^{k-1}e^{t_0 X}}{t - t_0} = \lim_{t \to t_0} \frac{M_X^{(k-1)}(t) - M_X^{(k-1)}(t_0)}{t - t_0} = \mathbb{E}X^k e^{t_0 X}. \tag{A.2}$$

Note that all the expectations in this equation are well defined (finite) as a result of part 1. Applying Taylor's formula (Theorem A.1, p. 159) to the function $h_x(t) = x^{k-1}e^{tx}$ (where x is seen as a fixed constant), we obtain

$$\frac{X^{k-1}e^{tX} - X^{k-1}e^{t_0 X}}{t - t_0} = \frac{h_X(t) - h_X(t_0)}{t - t_0} = h'_X(\xi) = X^k e^{\xi X}, \quad |\xi - t_0| \leq |t - t_0|.$$

Note that since ξ depends on both t and X, it's in fact a random variable. Similarly,

$$\frac{X^{k-1}e^{tX} - X^{k-1}e^{t_0 X}}{t - t_0} - X^k e^{t_0 X} = X^k e^{\xi X} - X^k e^{t_0 X}$$

$$= X^{k+1}e^{\xi' X}(\xi - t_0), \quad |\xi' - t_0| \leq |\xi - t_0|.$$

We must thus show that the expectation on the right-hand side tends to zero as $t \to t_0$. Since $|\xi - t_0| \leq |t - t_0|$, it suffices to bound $\mathbb{E}X^{k+1}e^{\xi' X}$ uniformly in t. Let $\delta > 0$ be such that $[t_0 - 2\delta, t_0 + 2\delta] \subset I$. Suppose without loss of generality that $|t - t_0| < \delta$. It follows that $t_0 - \delta \leq \xi \leq t_0 + \delta$ and we can use the same approach as before to write:

$$|X|^{k+1}e^{\xi' X} = X^{k+1}e^{\xi' X}\mathbb{1}\{X \geq 0\} + (-X)^{k+1}e^{\xi' X}\mathbb{1}\{X < 0\}$$

$$\leq X^{k+1}e^{(t_0+\delta)X}\mathbb{1}\{X \geq 0\} + (-X)^{k+1}e^{(t_0-\delta)X}\mathbb{1}\{X < 0\}$$

$$= e^{(t_0+2\delta)X}u_{k+1,\delta}(X)\mathbb{1}\{X \geq 0\} + e^{(t_0-2\delta)X}u_{k+1,\delta}(-X)\mathbb{1}\{X < 0\}.$$

It follows that

$$\mathbb{E}|X|^{k+1}e^{\xi'X} \leq C_{k+1,\delta}M_X(t_0 + 2\delta) + C_{k+1,\delta}M_X(t_0 - 2\delta) < \infty,$$

since $t_0 \pm 2\delta \in I$ and $C_{k+1,\delta} < \infty$. Hence

$$\mathbb{E}\left|\frac{X^{k-1}e^{tX} - X^{k-1}e^{t_0X}}{t - t_0} - X^k e^{t_0X}\right| \leq C_{k+1,\delta}[M_X(t_0 + 2\delta)$$

$$+ M_X(t_0 - 2\delta)]|t - t_0| \to 0, \quad t \to t_0.$$

Consequently, Eq. (A.2) holds true (since the term on the right-hand side of (A.2) is finite), which translates to

$$M_X^{(k)}(t_0) = \mathbb{E}X^k e^{t_0X} \qquad \forall t_0 \in I.$$

The recurrence thus holds true, which establishes 2 and 3. To complete the proof, observe that when $\{0\} \subset I$, part 4 follows directly from parts 1 and 3.   □

A further important property of the MGF is that, provided that M_X exists on an open interval containing zero, it *uniquely* determines the distribution of X:

Proposition A.9 (Characterisation Property of the MGF) *Let X and Y be two random variables taking values in $\mathbb{R}$, and let F_X and F_Y be their respective distributions. Let $M_X, M_Y : \mathbb{R} \to \mathbb{R}$ be their MGFs. If there exists an open interval I containing zero, such that $M_X(t) < \infty$ and $M_Y(t) < \infty$ for all $t \in I$, then*

$$F_X = F_Y \iff M_X = M_Y.$$

We will not prove this result in its full generality, as this would require either notions related to Laplace transforms, or to characteristic functions (see, e.g., Billingsley [2], Sect. 30). We will only give a proof for the special case of non-negative random variables (following a saddlepoint argument of Dalang and Conus [8]). This suffices to cover the situations where we will use the theorem in this text.

Proof of Proposition A.9, Assuming $X, Y \geq 0$ We first consider the case of continuous random variables, and focus on the random variable $X \geq 0$. Since $X \geq 0$, it follows that $M_X(t) < \infty$ for all $t < 0$. Combining this fact with our assumption, means that there exists a $\delta > 0$ such that $M_X(t) < \infty$ for all $t < \delta$. By Proposition A.8, we now know that $\frac{d^k}{dt^k}M_X$ exists for all k and all $t < \delta$. Our strategy will be to express F_X as a function of the derivatives of M_X. More specifically, define the function $G_X(t, x) : [0, \infty)^2 \to \mathbb{R}$ as

$$G_X(t, x) = \sum_{k=0}^{\lfloor tx \rfloor} \frac{t^k}{k!} \frac{d^k M_X}{dt^k}(-t),$$

where $\lfloor z \rfloor$ is the largest integer less than or equal to z. We will show that for any given $x \geq 0$,

$$\lim_{t \to \infty} G_X(t, x) = F_X(x).$$

Fix $x \geq 0$. Proposition A.8 shows that, for all $k \geq 0$,

$$\frac{d^k}{dt^k} M_X(t) = \mathbb{E}\left[X^k e^{tX}\right] = \int_0^\infty x^k e^{tx} f_X(x) dx,$$

where the last integral is over $[0, \infty)$ by non-negativity of X. Thus G may be re-expressed as

$$G_X(t, x) = \sum_{k=0}^{\lfloor tx \rfloor} \frac{t^k}{k!} \int_0^{+\infty} y^k e^{-ty} f_X(y) dy = \int_0^{+\infty} \underbrace{\left(\sum_{k=0}^{\lfloor tx \rfloor} \frac{t^k}{k!} y^k e^{-ty}\right)}_{=\varphi_t(x,y)} f_X(y) dy,$$

where $\varphi_t(x, y) = \mathbb{P}[W_{t,y} \leq tx]$ for $W_{t,x} \sim Poisson(ty)$. Consequently, when $y > x$, Chebyshev's inequality (Lemma A.4, p. 159) implies that

$$0 \leq \varphi_t(x, y) = \mathbb{P}[W_{t,y} \leq tx] = \mathbb{P}[W_{t,y} - ty \leq t(x - y)]$$
$$\leq \mathbb{P}[|W_{t,y} - ty| \geq t(y - x)]$$
$$\leq \frac{\text{Var}[W_{t,y}]}{t^2(y - x)^2} = \frac{y}{t(y - x)^2}.$$

Similarly, in the case $y < x$, we have

$$0 \leq 1 - \varphi_t(x, y) = \mathbb{P}[W_{t,y} > tx] = \mathbb{P}[W_{t,y} - ty > t(x - y)]$$
$$\leq \mathbb{P}[|W_{t,y} - ty| > t(x - y)]$$
$$\leq \frac{\text{Var}[W_{t,y}]}{t^2(x - y)^2} = \frac{y}{t(x - y)^2}.$$

Now let $\epsilon > 0$. Choose $h > 0$ sufficiently small so that $F_X(x + h) - F_X(x) < \epsilon/3$ and $F_X(x) - F_X(x - h) < \epsilon/3$ (such a choice is ensured by continuity of F_X). Then choose $t > 0$ sufficiently large so that $t > 6x/\epsilon h^2$. We have

$$|G_X(t, x) - F_X(x)| = \left| \int_0^{+\infty} \varphi_t(x, y) f_X(y) dy - \int_0^x f_X(y) dy \right|$$

$$= \left| \int_0^{x-h} (\varphi_t(x, y) - 1) f_X(y) dy + \int_{x-h}^x (\varphi_t(x, y) - 1) f_X(y) dy \right.$$

$$\left. + \int_x^{x+h} \varphi_t(x, y) f_X(y) dy + \int_{x+h}^\infty \varphi_t(x, y) f_X(y) dy \right|$$

$$\le \int_0^{x-h} |\varphi_t(x, y) - 1| f_X(y) dy + \int_{x-h}^x |\varphi_t(x, y) - 1| f_X(y) dy$$

$$+ \int_x^{x+h} |\varphi_t(x, y)| f_X(y) dy + \int_{x+h}^\infty |\varphi_t(x, y)| f_X(y) dy.$$

Let us consider the terms on the right-hand side one at a time, and bound them suitably (note that if $x = 0$, we only need to consider the last two integrals). We have

$$\int_0^{x-h} |\varphi_t(x, y) - 1| f_X(y) dy \le \frac{1}{t} \int_0^{x-h} \frac{y}{(x - y)^2} f_X(y) dy$$

$$\le \frac{x - h}{th^2} \int_0^{x-h} f_X(y) dy \le \frac{x - h}{th^2},$$

by our earlier calculation. Similarly,

$$\int_{x+h}^\infty |\varphi_t(x, y)| f_X(y) dy \le \frac{x + h}{th^2}.$$

Furthermore, $|\varphi_t(x, y) - 1| \le 1$ and $|\varphi_t(x, y)| \le 1$ for all $x, y \ge 0$, so that

$$\int_{x-h}^x |\varphi_t(x, y) - 1| f_X(y) dy \le \int_{x-h}^x f_X(y) dy = F_X(x) - F_X(x - h)$$

and

$$\int_x^{x+h} |\varphi_t(x, y)| f_X(y) dy \le \int_x^{x+h} f_X(y) dy = F_X(x + h) - F_X(x).$$

In summary, we have shown that for all $t > \frac{6x}{\epsilon h^2}$,

$$
\begin{aligned}
|G_X(t, x) - F_X(x)| &\leq \frac{x - h}{th^2} + [F_X(x) - F_X(x - h)] \\
&\quad + [F_X(x + h) - F_X(x)] + \frac{x + h}{th^2} \\
&= [F_X(x) - F_X(x - h)] + [F_X(x + h) - F_X(x)] + \frac{2x}{th^2} \\
&= \frac{\epsilon}{3} + \frac{\epsilon}{3} + \frac{\epsilon}{3} = \epsilon.
\end{aligned}
$$

In other words, we have shown that $|G_X(t, x) - F_X(x)| < \epsilon$ for any $\epsilon > 0$ and t sufficiently large, which proves that $\lim_{t \to \infty} G_X(t, x) = F_X(x)$. The exact same arguments show that $\lim_{t \to \infty} G_Y(t, x) = F_Y(x)$, where $G_Y(t, x)$ is defined in analogous fashion as $G_X(t, x)$. But $G_X = G_Y$ since $M_X = M_Y$, which proves that $F_X = F_Y$ and completes the proof in the case when the random variables X, Y are continuous. For the discrete case, we follow the exact same argument, replacing integrals by sums, and proving that $\lim_{t \to \infty} G_X(t, x) = F_X(x)$ for all continuity points x of $F_X(x)$. For discontinuity points, we then simply use the right-continuity of F_X. The proof is now complete.   □

The next lemma is useful when trying to establish the distribution of a sum or independent random variables.

Lemma A.10 (Sums and MGFs) *Let X and Y be two independent random variables taking values in $\mathbb{R}$, and let $Z = X + Y$. If $M_X(t) < \infty$ and $M_Y(t) < \infty$ for all t in an open interval I, then $M_Z(t) < \infty$ for all $t \in I$ and*

$$
M_Z(t) = M_X(t) M_Y(t).
$$

Proof By independence, we may write

$$
\begin{aligned}
\infty > M_X(t) M_Y(t) &= \mathbb{E}[e^{tX}] \mathbb{E}[e^{tY}] = \mathbb{E}[e^{tX} e^{tY}] \\
&= \mathbb{E}[\exp\{t(X + Y)\}] = M_Z(t), \qquad t \in I.
\end{aligned}
$$

□

A.7   Continuous Mapping and Slutsky's Theorem

In order to prove these two results, we will first need a couple of results regarding distribution functions and their convergence.

Lemma A.11 *Let F be a cumulative distribution function. Then F has at most countably many discontinuities.*

Proof Let D_F be the set of discontinuity points of F. Given any $x \in D_F$, we have

$$\lim_{\epsilon \downarrow 0} F(x - \epsilon) < \lim_{\epsilon \downarrow 0} F(x + \epsilon)$$

since F is non-decreasing. It follows that there exists a rational number $q(x)$ such that

$$\lim_{\epsilon \downarrow 0} F(x - \epsilon) < q(x) < \lim_{\epsilon \downarrow 0} F(x + \epsilon), \qquad \forall x \in D_F.$$

Furthermore, whenever $x_1 < x_2$ (so that we may write $x_2 = x_1 + \delta$, for some $\delta > 0$), the fact that F is non-decreasing implies that

$$q(x_1) < \lim_{\epsilon \downarrow 0} F(x_1 + \epsilon) \le F(x_1 + \delta/2) = F(x_2 - \delta/2) \le \lim_{\epsilon \downarrow 0} F(x - \epsilon) < q(x_2).$$

Summarising, we have constructed an injection $q : D_F \to \mathbb{Q}$, and thus D_F must be countable. □

Lemma A.12 *Given a sequence of random variables $X, X_1, X_2, \ldots$, the following two statements are equivalent:*
1. $X_n \overset{d}{\to} X$.
2. For all closed subsets $C \subseteq \mathbb{R}$, one has

$$\limsup_{n \to \infty} \mathbb{P}(X_n \in C) \le P(X \in C).$$

Proof Assume first that (2) holds true, so that for $C_1 = (-\infty, a]$ and $C_2 = [a, \infty)$, we have

$$\mathbb{P}(X < a) = 1 - \mathbb{P}(X \ge a) \le 1 - \limsup_{n \to \infty} \mathbb{P}(X_n \ge a) = \liminf_{n \to \infty} \mathbb{P}(X_n < a)$$

$$\le \liminf_{n \to \infty} \mathbb{P}(X_n \le a) \le \limsup_{n \to \infty} \mathbb{P}(X_n \le a) \le \mathbb{P}(X \le a).$$

If a is a continuity point of the distribution function of X, it must be that $\mathbb{P}(X < a) = \mathbb{P}(X \leq a)$ and so $\mathbb{P}(X_n \leq a) \to \mathbb{P}(X \leq a)$. This establishes that $X_n \xrightarrow{d} X$.

To prove the converse, assume initially that $C = [a, b]$, where $-\infty < a \leq b < \infty$. There exist sequences $0 \leq \epsilon_k \searrow 0, 0 \leq \delta_k \searrow 0$ such that $F(x) = \mathbb{P}(X \leq x)$ is continuous at the points $a - \delta_k$ and $b + \epsilon_k$ for all k (Lemma A.11). Consequently,

$$\limsup_{n \to \infty} \mathbb{P}(X_n \in C) \leq \limsup_{n \to \infty} \mathbb{P}(a - \delta_k < X_n \leq b + \epsilon_k) = \limsup_{n \to \infty} \mathbb{P}(X_n \leq b + \epsilon_k)$$

$$-\mathbb{P}(X_n \leq a - \delta_k) = \mathbb{P}(X \leq b + \epsilon_k) - \mathbb{P}(X \leq a - \delta_k) = \mathbb{P}(a - \delta_k < X \leq b + \epsilon_k).$$

Letting $k \to \infty$, continuity from above of probability measures yields

$$\limsup_{n \to \infty} \mathbb{P}(X_n \in C) \leq \lim_{k \to \infty} \mathbb{P}(a - \delta_k < X \leq b + \epsilon_k)$$

$$= \mathbb{P}\left(\bigcap_{k=1}^{\infty} \{a - \delta_k < X \leq b + \epsilon_k\} \right) = \mathbb{P}(X \in C).$$

If $a = -\infty$ or $b = \infty$, the statement can be shown to be true by a similar argument. Thus (2) is true when C is an interval.

If $C = \cup C_k$ is the countable union of (potentially infinitely many) closed disjoint intervals, the subadditivity of limit superior yields

$$\limsup_{n \to \infty} \mathbb{P}(X_n \in C) = \limsup_{n \to \infty} \sum_{k=1}^{\infty} \mathbb{P}(X_n \in C_k) \leq \sum_{k=1}^{\infty} \limsup_{n \to \infty} \mathbb{P}(X_n \in C_k)$$

$$\leq \sum_{k=1}^{\infty} \mathbb{P}(X \in C_k) = \mathbb{P}(X \in C).$$

Suppose now that $C = \cap C_k$, where each C_k is a disjoint union of countably many closed intervals, and $C_{k+1} \subseteq C_k$ for all k. Following the same course as in the first part of the proof,

$$\limsup_{n \to \infty} \mathbb{P}(X_n \in C) \leq \limsup_{n \to \infty} \mathbb{P}(X_n \subset C_k) \leq \mathbb{P}(X \subset C_k) \searrow \mathbb{P}(X \subset C), \quad k \to \infty.$$

To complete the proof, thus, it suffices to show that any closed set $C \subseteq \mathbb{R}$ can be written in this form.

For every k, divide $\mathbb{R}$ into closed intervals of length 2^{-k}, that is $I_j^{(k)} = 2^{-k}[j, j + 1]$. Let C_k be the union of those intervals $\{I_j^{(k)}\}$ that have a non-empty intersection with C:

$$C_k = \bigcup_{j \in \mathbb{Z}: I_j^{(k)} \cap C \neq \emptyset} I_j^{(k)}.$$

It is clear that C_k is the countable union of countably many closed intervals, and that $C_k \supseteq C$. If $x \notin C$, there exists an interval I such that $C \cap I = \emptyset$ that contains x. For k such that $2^{-k} < m(I)/2$ it follows that $x \notin C_k$. We may thus conclude that $C = \cap C_k$. The fact that C_k is closed follows by a similar reasoning, but we can argue differently: let $x_n \in C_k$ be a sequence converging to x. There must exist an M such that the sequence is contained in $C_k \cap [-M, M]$. This last set is closed, as it is the union of finitely many closed intervals. Hence $x \in C_k \cap [-M, M]$ and so C_k is closed.

It remains to show that $C_{k+1} \subseteq C_k$. Let $x \in C_{k+1}$. There exists $j \in \mathbb{Z}$ such that $x \in I_j^{(k+1)} \subseteq C_{k+1}$. Or, $I_j^{(k+1)} \subset I_{\lfloor j/2 \rfloor}^{(k)}$, and thus this last set has a non-empty intersection with C. It follows that $x \in I_{\lfloor j/2 \rfloor}^{(k)} \subseteq C_k$, and the proof is complete. □

Proof of the Continuous Mapping Theorem (Theorem 2.25, p. 57) By Lemma A.12, it suffices to prove that $X_n \overset{d}{\to} X$ implies $\limsup\limits_{n \to \infty} \mathbb{P}[g(X_n) \le y] \le \mathbb{P}[g(X) \in C]$ for all closed $C \subseteq \mathbb{R}$. To this aim, let $C \subseteq \mathbb{R}$ be an arbitrary closed set, let

$$A = \{x \in \mathbb{R} : g(x) \in C\}$$

be the inverse image of C via g, and let $\overline{A}$ denote the closure of A. If D_g is the set of discontinuities of g, we may write

$$\overline{A} = \left\{\overline{A} \cap D_g\right\} \cup \left\{\overline{A} \cap D_g^c\right\} \subseteq D_g \cup \left\{\overline{A} \cap D_g^c\right\}.$$

Now if $x \in \overline{A} \cap D_g^c$, then there exists a sequence $\{x_k\} \subset A$ such that $\lim_{k \to \infty} x_k = x$ (by definition of the closure, $\overline{A}$). Furthermore, it holds that $g(x) = \lim_{k \to \infty} g(x_k) \in C$, because $x \in D_g^c$ also. Consequently $x \in A$, and we have proven that $\overline{A} \cap D_g^c \subseteq A$.

Summarising, we have

$$\overline{A} \subseteq A_y \cup D_g. \tag{A.3}$$

We now exploit this inclusion in order to write

$$\mathbb{P}[g(X_n) \subset C] = \mathbb{P}[X_n \in A] \le \mathbb{P}[X_n \in \overline{A}].$$

But,

$$\limsup_{n\to\infty} \mathbb{P}[X_n \in \overline{A}] \leq \mathbb{P}[X \in \overline{A}] \quad [\text{using } X_n \overset{d}{\to} X, \text{ combined with Lemma A.12}]$$

$$\leq \mathbb{P}[X \in A \cup D_g] \quad [\text{by (A.3)}]$$

$$\leq \mathbb{P}[X \in A] + \underbrace{\mathbb{P}[X \in D_g]}_{=0}$$

$$= \mathbb{P}[g(X) \in C].$$

It follows that $\limsup_{n\to\infty} \mathbb{P}[g(X_n) \in C] \leq \mathbb{P}[g(X) \in C]$ and our proof is complete.

□

Proof of Slutsky's Theorem (Theorem 2.26, p. 57) For the first part, assume that $X_n \overset{d}{\to} X$ and $Y_n \overset{p}{\to} c$. We may assume without loss of generality that $c = 0$. Let x be a continuity point of F_X. We have

$$\mathbb{P}[X_n + Y_n \leq x] = \mathbb{P}[X_n + Y_n \leq x, |Y_n| \leq \epsilon] + \mathbb{P}[X_n + Y_n \leq x, |Y_n| > \epsilon]$$

$$\leq \mathbb{P}[X_n \leq x + \epsilon] + \mathbb{P}[|Y_n| > \epsilon]$$

because $\{X_n + Y_n \leq x \,\&\, |Y_n| \leq \epsilon\}$ implies that $\{X_n \leq x + \epsilon\}$. Similarly, we may obtain the inequality

$$\mathbb{P}[X_n \leq x - \epsilon] \leq \mathbb{P}[X_n + Y_n \leq x] + \mathbb{P}[|Y_n| > \epsilon].$$

Rearranging and collecting terms yields:

$$\mathbb{P}[X_n \leq x - \epsilon] - \mathbb{P}[|Y_n| > \epsilon] \leq \mathbb{P}[X_n + Y_n \leq x] \leq \mathbb{P}[X_n \leq x + \epsilon] + \mathbb{P}[|Y_n| > \epsilon]$$

$$\lim_{n\to\infty} \mathbb{P}[X_n \leq x - \epsilon] - 0 \leq \lim_{n\to\infty} \mathbb{P}[X_n + Y_n \leq x] \leq \lim_{n\to\infty} \mathbb{P}[X_n \leq x + \epsilon] + 0$$

By Lemma A.11, we may find a sequence $0 < \epsilon_k \downarrow 0$ such that $x + \epsilon_k$ is a continuity point, for all k. Replacing ϵ by ϵ_k gives

$$F_X(x - \epsilon_k) \leq \lim_{n\to\infty} \mathbb{P}[X_n + Y_n \leq x] \leq F_X(x + \epsilon_k).$$

Since x is a continuity point of F_X, letting $k \to \infty$ establishes $X_n + Y_n \overset{d}{\to} X$.

To prove the second part, let $Z_n = Y_n - c$, so that $Z_n \overset{p}{\to} 0$. Thus, if we can show $X_n Z_n \overset{d}{\to} 0$, then the conclusion follows by first part of the theorem, which is already proven. Let $\epsilon > 0$ and $M_k \uparrow \infty$ be positive sequence such that ϵM_k is a continuity point of $F_{|X|}$ for all k (this choice is feasible by Lemma A.11). Note also that $|X_n| \overset{d}{\to} |X|$ by the continuous mapping theorem (Theorem 2.25, 57). Combining these ingredients yields:

$$\mathbb{P}[|X_n Z_n| > \epsilon] \leq \mathbb{P}[|X_n Z_n| > \epsilon, |Z_n| \leq 1/M_k] + \mathbb{P}[|Z_n| \geq 1/M_k]$$
$$\leq \mathbb{P}[|X_n| > \epsilon M_k] + \mathbb{P}[|Z_n| \geq 1/M_k]$$
$$\leq 1 - \mathbb{P}[|X_n| \leq \epsilon M_k] + \mathbb{P}[|Z_n| \geq 1/M_k]$$
$$\implies \lim_{n \to \infty} \mathbb{P}[|X_n Z_n| > \epsilon] \leq \mathbb{P}[|X| > \epsilon M_k].$$

The right-hand side can be made arbitrarily small by choosing k sufficiently large. Thus $Z_n X_n \overset{p}{\to} 0$. Since $X_n Y_n = Z_n X_n + c X_n$, we use the first part of the theorem (already proven) to conclude that $X_n Y_n \overset{p}{\to} 0$. $\qquad\square$

A.8 On the Proof of the Central Limit Theorem

The standard proof of the central limit theorem makes use of the *characteristic function*, and thus involves notions from complex analysis, and more specifically the Lévy continuity theorem (see, e.g., Billingsley [2], Sect. 29). Since the latter result is beyond the scope of this text, we will provide an elementary proof here due to Lindeberg [17] (as presented in Dalang [7]), that is based on stronger assumptions, namely existence of a third absolute moment.[5],[6]

We first need three intermediate results. In what follows, $C_b^3(\mathbb{R})$ denotes the set of all thrice continuously differentiable bounded functions $\mathbb{R} \to \mathbb{R}$, that are bounded, and whose first three derivatives are also bounded.

Lemma A.13 *Let Z be a continuous random variable, and $\{Z_n\}_{n \geq 1}$ a sequence of random variables such that*

$$\mathbb{E}[g(Z_n)] \overset{n \to \infty}{\longrightarrow} \mathbb{E}[g(Z)]$$

for all $g \in C_b^3(\mathbb{R})$. Then

$$F_{Z_n}(x) \overset{n \to \infty}{\longrightarrow} F_Z(x), \qquad \forall x \in \mathbb{R}.$$

---

[5]As a matter of fact, even this weaker version of the theorem would suffice for the asymptotic results presented in this text: these require the sufficient statistic of an exponential family to satisfy the central limit theorem (as Corollary 2.24, p. 56), and the latter statistic will have finite moments of all orders (see Eq. (2.11), p. 51, in the proof of Proposition 2.11).

[6]The same method of proof can be "upgraded" to work under only second moment assumptions, assuming knowledge of measure theory, in particular the monotone convergence theorem (Dalang [7]).

Proof Let $x \in \mathbb{R}$ and $k \geq 1$ be given. Note that we may always choose a function $g_k \in C_b^3(\mathbb{R})$ that satisfies the envelope relation

$$\mathbf{1}\{z \in (-\infty, x]\} \leq g_k(z) \leq \mathbf{1}\{z \in (-\infty, x + 1/k]\}. \tag{A.4}$$

Then, for all $n \geq 1$,

$$F_{Z_n}(x) = \mathbb{P}[Z_n \leq x] = \mathbb{E}[\mathbf{1}\{z \in (-\infty, x]\}] \leq \mathbb{E}[g_k(Z_n)],$$

and hence by our assumption we have

$$\limsup_{n \to \infty} F_{Z_n}(x) \leq \lim_{n \to \infty} \mathbb{E}[g_k(Z_n)] = \mathbb{E}[g_k(Z)]$$

$$\leq \mathbb{E}[\mathbf{1}\{z \in (-\infty, x + 1/k\}] = F_Z(x + 1/k).$$

The same type of argument shows that $\liminf_{n \to \infty} F_{Z_n}(x) \geq F_Z(x - 1/k)$. Since the choice of k was arbitrary, and since F_Z is everywhere continuous, we have that $F_{Z_n}(x) \xrightarrow{n \to \infty} F_Z(x)$, completing the proof. $\qquad \square$

Lemma A.14 *Let $g \in C_b^3(\mathbb{R})$, and let $\sup_{x \in \mathbb{R}} |g'''(x)| = C < \infty$. Let (Y, Z) be independent random variables such that $\mathbb{E}[Y] = \mathbb{E}[Z]$, and $\mathbb{E}[Y^2] = \mathbb{E}[Z^2]$. If X is independent of Y and Z, we have*

$$\left| \mathbb{E}[g(X + Y) - g(X + Z)] \right| \leq \frac{C}{6} \left(\mathbb{E}|Y|^3 + \mathbb{E}|Z|^3 \right).$$

Proof Taylor's theorem (Theorem A.1, p. 159) yields that

$$g(x + y) = g(x) + yg'(x) + \frac{1}{2}y^2 g''(x) + \frac{1}{6}y^3 g'''(u),$$

where u lies between x and $x + y$. It follows now by independence that

$$\mathbb{E}[g(X + Y)] = \mathbb{E}[g(X)] + \mathbb{E}[Y]\mathbb{E}[g'(X)] + \frac{1}{2}\mathbb{E}[Y^2]\mathbb{E}[g''(X)] + \frac{1}{6}\mathbb{E}[Y^3 g'''(U)]$$

$$\mathbb{E}[g(X + Z)] = \mathbb{E}[g(X)] + \mathbb{E}[Z]\mathbb{E}[g'(X)] + \frac{1}{2}\mathbb{E}[Z^2]\mathbb{E}[g''(X)] + \frac{1}{6}\mathbb{E}[Z^3 g'''(V)]$$

for a random variable U that lies between X and $X + Y$ almost surely, and a random variable V that lies between X and $X + Z$ almost surely. Consequently, our assumptions yield that

$$\left| \mathbb{E}[g(X+Y) - g(X+Z)] \right| = \left| \frac{1}{6}\mathbb{E}[Y^3 g'''(U)] - \frac{1}{6}\mathbb{E}[Z^3 g'''(V)] \right|$$

$$\leq \frac{1}{6}\mathbb{E}\left| Y^3 g'''(U) \right| + \frac{1}{6}\mathbb{E}\left| Z^3 g'''(V) \right|$$

$$\leq \frac{C}{6}\left(\mathbb{E}|Y|^3 + \mathbb{E}|Z|^3 \right).$$

$\square$

Lemma A.15 *Let $\{\tilde{Y}_n\}_{n\geq 1}$ be a sequence of iid random variables such that $\mathbb{E}|\tilde{Y}_1|^3 < \infty$, $\mathbb{E}[\tilde{Y}_1^2] = 1$, and $\mathbb{E}[\tilde{Y}_1] = 0$. If $g \in C_b^3(\mathbb{R})$, then it holds that*

$$\mathbb{E}\left[g\left(\frac{\sum_{i=1}^n \tilde{Y}_i}{\sqrt{n}} \right) \right] \overset{n\to\infty}{\longrightarrow} \mathbb{E}\left[g(\tilde{Z}) \right],$$

where $\tilde{Z} \sim N(0,1)$.

Proof Let $g \in C_b^3(\mathbb{R})$, and $n \geq 1$. Let $\{\tilde{Z}_i\}_{i=1}^n \overset{\text{iid}}{\sim} N(0,1)$ (independent of the $\{\tilde{Y}_i\}$) and define

$$Y_i = \tilde{Y}_i / \sqrt{n} \quad \& \quad Z_i = \tilde{Z}_i / \sqrt{n}.$$

Since $\{\tilde{Z}_i\}_{i=1}^n \overset{\text{iid}}{\sim} N(0, 1/n)$, it follows that $\sum_{i=1}^n Z_i \sim N(0,1)$ (by Corollary 1.35, p. 25). It thus suffices to show that

$$\left| \mathbb{E}[g(Y_1 + \cdots Y_n)] - \mathbb{E}[g(Z_1 + \cdots Z_n)] \right| \leq \frac{C}{6} \frac{\mathbb{E}[|\tilde{Y}_1|^3] + \mathbb{E}[|\tilde{Z}_1|^3]}{\sqrt{n}} \tag{A.5}$$

for $C = \sup_{x\in\mathbb{R}} |g'''(x)| < \infty$. Define

$$U_i = Y_1 + \cdots + Y_{i-1} + Y_i + Z_{i+1} + \cdots + Z_n$$

$$V_i = Y_1 + \cdots + Y_{i-1} + 0 + Z_{i+1} + \cdots + Z_n$$

and observe that these satisfy

$$U_i = V_i + Y_i \quad \& \quad U_{i-1} = V_i + Z_i$$

so that we may re-write the left-hand side of Eq. (A.5) as

$$\mathbb{E}[g(U_n)] - \mathbb{E}[g(U_0)] = \sum_{i=1}^{n} \left(\mathbb{E}[g(U_i)] - \mathbb{E}[g(U_{i-1})] \right)$$

$$= \sum_{i=1}^{n} \left(\mathbb{E}[g(V_i + Y_i)] - \mathbb{E}[g(V_i + Z_i)] \right).$$

We now use Lemma A.14 to bound the last expression by

$$\sum_{i=1}^{n} \frac{C}{6} \left(\mathbb{E}[|Y_i|^3] - \mathbb{E}[|Z_i|^3] \right) = n \frac{C}{6} n^{-3/2} \left(\mathbb{E}[|\tilde{Y}_1|^3] + \mathbb{E}[|\tilde{Z}_1|^3] \right)$$

thus establishing the validity of inequality A.5, and completing the proof. □

Theorem A.16 (Third Moment Central Limit Theorem) *Let $Y_1, \ldots, Y_n$ be iid random variables such that $\mathbb{E}[Y_i] = \mu < \infty$, $\mathrm{Var}[Y_i] = \sigma^2$, and $\mathbb{E}|Y_i|^3 < \infty$. Let $\overline{Y}_n = \frac{1}{n} \sum_{i=1}^{n} Y_i$. Then,*

$$\sqrt{n}(\overline{Y}_n - \mu) \xrightarrow{d} N(0, \sigma^2).$$

Proof The random variables $\tilde{Y}_i = \frac{Y_i - \mu}{\sigma}$ satisfy the conditions of Lemma A.15. Thus, if we define

$$Z_n := \frac{\tilde{Y}_1 + \ldots + \tilde{Y}_n}{\sqrt{n}} = \frac{\sqrt{n}(\overline{Y}_n - \mu)}{\sigma},$$

we must have

$$\mathbb{E}[g(Z_n)] \xrightarrow{n \to \infty} \mathbb{E}\left[g(Z) \right], \qquad \forall g \in C_b^3(\mathbb{R}),$$

for $Z \sim N(0, 1)$. Lemma A.13 now implies that $F_{Z_n}(x) \xrightarrow{n \to \infty} F_Z(x)$ for all $x \in \mathbb{R}$, and so $\sigma Z_n = \sqrt{n}(\overline{Y}_n - \mu) \xrightarrow{d} N(0, \sigma^2)$. □

Bibliography

1. Bickel, P. J., & Doksum, K. A. (2001). *Mathematical statistics: Basic ideas and selected topics.* Upper Saddle River: Prentice Hall.
2. Billingsley, P. (1986). *Probability and measure.* New York: Wiley.
3. Blitzstein, J. K., & Hwang, J. (2015). *Introduction to probability.* London: Chapman & Hall/CRC.
4. Casella, G., & Berger, R. L. (2002). *Statistical inference.* Pacific Grove: Duxbury Press.
5. Corwin, L. J., & Szczarba, R. H. (1982). *Multivariable calculus.* New York: Marcel Dekker.
6. Cox, D. R., & Hinkley, D. V. (1979). *Theoretical statistics.* London: Chapman & Hall/CRC.
7. Dalang, R. C. (2006). Une démonstration élémentaire du théorème central limite. *Elemente der Mathematik, 61*(2), 65–73.
8. Dalang, R. C., & Conus, D. (2008). *Introduction à la théorie des probabilités.* Lausanne: Presses Polytechniques et Universitaires Romandes.
9. Davison, A. C. (2003). *Statistical models.* Cambridge: Cambridge University Press.
10. Durrett, R. (1996). *Probability: Theory and examples.* Pacific Grove:: Duxbury Press.
11. Grimmett, G., & Welsh, D. (2014). *Probability: An introduction.* Oxford: Oxford University Press.
12. Hogg, R. V., & Craig, A. T. (1970). *Introduction to mathematical statistics.* New York: Macmillan.
13. Hogg, R. V., & Tanis, E. A. (2000). *Probability and statistical inference.* Upper Saddle River: Prentice Hall.
14. Knight, K. (2000). *Mathematical statistics.* Boca Raton: Chapman & Hall/CRC.
15. Lehmann, E. L., & Casella, G. (2003). *Theory of point estimation.* New York: Springer.
16. Lehmann, E. L., & Romano, J. P. (2008). *Testing statistical hypotheses.* New York: Springer.
17. Lindeberg, J. (1922). Eine neue Herleitung des Exponentialgesetzes in der Wahrscheinlichkeitsrechnung. *Mathematische Zeitschrift, 15*, 211–225.
18. Pitman, J. (1993). *Probability.* New York: Springer.
19. Rice, J. A. (2006). *Mathematical statistics and data analysis.* Belmont: Duxbury Press.
20. Ross, S. M. (2010). *A first course in probability.* Upper Saddle River: Prentice Hall.
21. Rudin, W. (1976). *Principles of mathematical analysis.* New York: McGraw-Hill.
22. Schervish, M. J. (2010). *Theory of statistics.* New York: Springer.
23. Shao, J. (2008). *Mathematical statistics.* New York: Springer.
24. Silvey, S. D. (2003). *Statistical inference.* London: Chapman & Hall/CRC.
25. Wasserman, L. (2004). *All of statistics: A concise course in statistical inference.* New York: Springer.
26. Young, G. A., & Smith, R. L. (2005). *Essentials of statistical inference.* Cambridge: Cambridge University Press.

© Springer International Publishing Switzerland 2016
V.M. Panaretos, *Statistics for Mathematicians*, Compact Textbooks in Mathematics,
DOI 10.1007/978-3-319-28341-8

Printed in the United States
By Bookmasters